Workshop on Dimensions of Meaning: Distributional and Curated Semantics (DistCurate 2022)

Online
14 July 2022

ISBN: 978-1-7138-5629-0

Printed from e-media with permission by:

Curran Associates, Inc.
57 Morehouse Lane
Red Hook, NY 12571

CURRAN ASSOCIATES INC.
proceedings
.com

Some format issues inherent in the e-media version may also appear in this print version.

Printed with permission by Curran Associates, Inc. (2022)

Association for Computational Linguistics
209 N. Eighth Street
Stroudsburg, Pennsylvania 18360

Phone: 1-570-476-8006
Fax: 1-570-476-0860

acl@aclweb.org

Additional copies of this publication are available from:

Curran Associates, Inc.
57 Morehouse Lane
Red Hook, NY 12571 USA
Phone: 845-758-0400
Fax: 845-758-2633
Email: curran@proceedings.com
Web: www.proceedings.com

DistCurate 2022

Workshop on Dimensions of Meaning: Distributional and Curated Semantics

Proceedings of the Workshop

July 14, 2022

Order copies of this and other ACL proceedings from:

Association for Computational Linguistics (ACL)
209 N. Eighth Street
Stroudsburg, PA 18360
USA
Tel: +1-570-476-8006
Fax: +1-570-476-0860
acl@aclweb.org

Organizing Committee

Organizers

Collin F. Baker, International Computer Science Institute
Michael Ellsworth, International Computer Science Institute
Miriam R. L. Petruck, International Computer Science Institute

Program Committee

Program Committee

Omri Abend, Hebrew University of Jerusalem
Gerard de Melo, University of Potsdam
Katrin Erk, University of Texas at Austin
Annette Frank, Heidelberg University
Richard Futrell, University of California, Irvine
Christopher Potts, Stanford University
Michael Roth, Saarland Univerity
Nathan Schneider, Georgetown University

Keynote Talk: Lexical semantics in the time of large language models

Chrsitopher Potts
Stanford University

Abstract: Present-day language models provide contextual representations of words: a given element of the model's vocabulary will generally be represented differently depending on the larger (linguistic and/or non-linguistic) context in which it appears. This is undoubtedly a key feature of the success of these models. What do these contextual representations mean for linguists working on lexical semantics? In this talk, I'll argue, first and foremost, that contextual representations are better aligned with linguists' conception of word meaning than any previous representation scheme in NLP, including symbolic approaches. I will then describe some ways in which linguists can use large pretrained language models to gain new insights into lexical meaning, and some ways in which language model development could be fruitfully informed by findings in linguistics. The overall take-away is that the turn toward contextual representations has created an exciting new space for collaboration between linguists and NLPers.

Bio: Christopher Potts is Professor and Chair of Linguistics and Professor (by courtesy) of Computer Science at Stanford, and a faculty member in the Stanford NLP Group and the Stanford AI Lab. His group uses computational methods to explore topics in emotion expression, context-dependent language use, systematicity and compositionality, and model interpretability. This research combines methods from linguistics, cognitive psychology, and computer science, in the service of both scientific discovery and technology development. He is the author of the 2005 book The Logic of Conventional Implicatures as well as numerous scholarly papers in computational and theoretical linguistics.

Table of Contents

A descriptive study of metaphors and frames in the multilingual shared annotation task

Maucha Andrade Gamonal
Laboratory for Experimentation in Translation
Federal University of Minas Gerais, Belo Horizonte, Brazil
mgamonal@ufmg.br

Abstract

This work assumes that languages are structured by semantic frames, which are schematic representations of concepts. Metaphors, on the other hand, are cognitive projections between domains, which are the result of our interaction in the world, through experiences, expectations and human biology itself. In this work, we use both semantic frames and metaphors in multilingual contrast (Brazilian Portuguese, English and German). The aim is to present a descriptive study of metaphors and frames in the multilingual shared annotation task of Multilingual FrameNet, a task which consisted of using frames from Berkeley FrameNet to annotate a parallel corpora. The result shows parameters for the metaphorical comparison considering those frames.

1 Introduction

Understanding human language requires activating cognitive models of socially shared knowledge. The linguistic units, whether lexical or constructional items, evoke possibilities of meaning defined in the context of linguistic-conceptual models, which are called semantic frames (Fillmore, 1982, 1985).

The computational implementation of frames was created at the end of the 20th century for the English language by the FrameNet project. The semantic frames methodology was later expanded to other languages, including Brazilian Portuguese (Torrent et al., 2022).

Motivated by the interest in frame comparison in a multilingual perspective, Multilingual FrameNet completed its first task, which consisted of the parallel corpora annotation of the TED talk "Do schools Kill Creativity?" (Robinson, 2016). This work tests the alignment of its linguistic databases with the frames defined for English. The goal is to establish the means of creating multilingual lexical resources as well as a semantically referenced machine translator (Torrent et al., 2018).

This article shows a comparative study of metaphors found in these linguistic annotations for Brazilian Portuguese, English and German. Bringing the results together, we have a set of metaphorical metadata in terms of metaphors and frames. The work brought together the theoretical contributions of Frame Semantics, created by Fillmore (1982), and the Conceptual Theory of Metaphor, compiled by Lakoff and Johnson (1999).

Both theories have computational applications: FrameNet (Ruppenhofer et al., 2016) and MetaNet (Dodge et al., 2015). The present work follows the methodological guidelines of FrameNet and uses MetaNet in order to explore the metaphors described in its network.

The text is organized as follows: section 2 presents Frame Semantics and the Conceptual Theory of Metaphor are presented; section 3 discusses semantic frames and translation studies; section 4 explains the multilingual annotation task; section 5 presents the parameterization of metaphorical metadata, and section 6 presents the summary and mentions some directions to take in a future work.

2 Background

2.1 Frame Semantics

Frame Semantics was created by Fillmore (1982, 1985). It assumes that the meaning of a linguistic unit, lexical or constructional, underlies a network of other units, which suggests the interactivity of meaning in a natural language. The term *frame* designates the socially shared linguistic-conceptual model that structures this knowledge representation.

> "By the term 'frame' I have in mind any system of concepts related in such a way that to understand any one of them you have to understand the whole structure in which it fits; when one of the things in such a structure is introduced in a text, or

Proceedings of the Workshop on Dimensions of Meaning: Distributional and Curated Semantics (DistCurate 2022), pages 1 - 7
July 14, 2022

into a conversation, all of the others are automatically available" (Fillmore, 1982, 111).

The study of semantic frames is a precursor of Cognitive Linguistics as a field. Different authors may use their own terminology and have their own specific purposes in using concepts such as 'domains' in Langacker (1987), 'Idealized Cognitive Models' and 'scripts' in Lakoff and Turner (2009) or may use frames to develop a concept such as the Mental Spaces theory in Fauconnier (1994).

(1) [The company $_{\text{Employer}}$] **HIRED** [him $_{\text{Employee}}$] [after going through a lengthy selection process $_{\text{Time}}$].

(2) [She $_{\text{Employee}}$] was **DISMISSED** [by the manager $_{\text{Employer}}$] [after fifteen years in the role $_{\text{Time}}$].

To understand the concept of frames, take the employment event as an example. In this situation, an employee and an employer begin an employment relationship, in which the employee will perform some predetermined activity for the employer, in exchange for payment. For a period of time, the employee remains employed, and the relationship ends when the employee quits the job, the employer dismisses him or the employee retires.

Sentences (1) and (2) show parts of the sequence of events in an employment activity. The Lexical Units *hire.v* and *fire.v* evoke `Hiring` and `Firing` frames, respectively. These frames are defined in terms of its participants, props and other conceptual roles, which are the semantic roles of the lexical units shown in bold above, called frame elements.

2.2 Conceptual Metaphor Theory

The Conceptual Metaphor Theory was proposed by Lakoff and Johnson (2008) with complementary contributions added later on Lakoff (2008); Lakoff and Turner (2009); Lakoff and Johnson (1999). Many studies show that metaphors are not limited to ornaments of speech or writing, but are the result of our interaction in the world, through experiences, expectations and human biology itself.

Lakoff (2012) takes into account the intrinsic relationship of metaphors with human biology itself, and discusses metaphor as evidence of embodied cognition. He suggests that neural mappings in the brain are related to metaphorical domain correspondences.

As a result of the metaphorical phenomenon, we understand things that are more abstract or subjective and less structured in terms of others that are more concrete, objective or more structured.

(3) And my contention is, all kids have tremendous talents. And we squander them, pretty ruthlessly. (TEDTalk)

In the example (3), the metaphor TALENT IS A RESOURCE is used. Through it, an element of the experiential domain of an attribute is understood as an entity or, more specifically, a finite resource.

A metaphorical projection presupposes a correspondence between domains: the source domain structures what is intended by the target. In this work, the relation between domains is explained through FrameNet frames. According to the FrameNet's annotation procedure, metaphors are marked by an extra annotation layer. If the metaphor is productive, it is indicated in the source domain. If it occurs at the level of the lexicon, the lexical unit will be in the target frame, the one that specifies the speaker's intention when producing an utterance (Ruppenhofer et al., 2016).

3 Frames and translation

Schäffner (2004, 2016) approaches the metaphorical phenomenon through a connection between Cognitive Linguistics and translation studies. She revisits the analytical methods of theorists such as Newmark (1981) and Toury (1995).

The Primacy of Frame Model of Translation hypothesis (Czulo, 2017) proposes the use of semantic frames as a descriptive basis for translational comparison. Czulo's hypothesis is based on one of the premises of Frame Semantics: when a frame is evoked, several others are automatically activated. As the author points out, metaphors are candidates for this frame co-activation process.

Theoretical and descriptive advances on the topic of frames and translation have been made. However, computational models and applications to automatically assess the translation of metaphors according to cognitive linguistics assumptions is an open topic that requires multidisciplinary research.

4 The multilingual annotation task

The first task of the Multilingual FrameNet initiative was to create a semantically refined linguistic analysis sample that would allow database alignment tests. FrameNets developed for different languages were responsible for annotating parallel corpora aligned at the sentence level. These annotations were produced using the text of a conference in the TED Talk model (Robinson, 2016)[1].

The analyses were carried out using the full-text annotation method, which consists of all semantic frame-evoking lexical units being submitted for annotation. The semantic frames used for the annotation task were those from Berkeley FrameNet, which highlights the initiative of developing computer applications for multilingual alignment purposes based on the Berkeley FrameNet database (Baker and Lorenzi, 2020).

Figure 1 shows an example of the annotation in English. The highlighted items are the Lexical Units under analysis. Each lexical unit evokes a frame, and the frame, in turn, brings a series of elements, the so-called Frame Elements, all named specifically in relation to the frame they belong to, as shown in the lower description of the respective figures.

In addition to the semantic analysis, the syntactic treatment is also included. It distinguishes the Grammatical Functions of Frame Elements as well as their Phrase Types. For this reason, the linguistic analysis of a framenet is commonly called a three-layer annotation. Although other layers may exist, the three essential ones are: essentially, Frame Element (FE), Grammatical Function (GF) and Phrasal Type (PT).

[1] https://www.ted.com/talks/sir_ken_robinson_do_schools_kill_creativity/transcript

5 The parameterization of metaphorical metadata

In this present work, fifty sentences from the corpus were analyzed. One of the procedures adopted was to start the comparative study using the metaphors in the Brazilian Portuguese text and, then, to interpret such translation choices in German and check the original text in English.

The result of this process is a set of analyses that indicate paths to a descriptive method of extracting semantic information. The parameters for the metaphorical comparison considering the frames evoked in the three languages were: total, directly related, indirectly related, and unrelated.

i Total: the frames are the same.

ii Directly related: the relation is direct. The frames are connected by one of the frame-to-frame relations.

iii Indirectly related: the relation is indirect. They are connected by one of the frame-to-frame relations, while expanding the network.

iv Unrelated: there is no relation between frames in the FrameNet database.

The analysis took into account the metaphorical projections in Brazilian Portuguese. As shown through the examples, a metaphor in Brazilian Portuguese may have resulted from different translation options of the original in English, and the correlation with the German version also occurred in different ways.

Ted_Creativity_corpus : doc_ted_en	
idSentence	Sentence
1043	What you have there is a person of extraordinary dedication who found a talent.
Sentences	

What you have there is a person of EXTRAORDINARY^{Desirability} [$_{Evaluee}$dedication] who found a talent.

What you have there is a [$_{Person}$person] of extraordinary dedication who found a talent.
PERSON^{People}

[$_{Entity}$What] [$_{Topical_entity}$you] HAVE^{Have_associated} [$_{Place}$there] is a person of extraordinary dedication who found a talent.

What you have there is [$_{Perceiver}$a person of extraordinary dedication] [$_{Perceiver}$who] Found^{Locating} [$_{Sought_entity}$a talent].

What you have there is a person of extraordinary dedication who found a TALENT^{Capability}.[$_{Entity}$CNI][$_{Event}$DNI]

© 2014, 2021 FrameNetBrasil Project

Figure 1: Semantic annotation of the multilingual task in English.

Table 1 summarizes the process of analyzing an example in which the match is total. The metaphorical sentence in Brazilian Portuguese has *perder.v* as a Lexical Unit which evokes the `Losing` frame in the FrameNet database. In German, the translation choice is *verloren.v*, evoking the same frame, and, in the original text, the speaker's lexical option was *lose.v*, evoking the same frame.

The parallel corpora contrasts each corpus to the original. In Table 1, both German and Brazilian Portuguese languages start from a metaphor and are translated through the same metaphor: ATTRIBUTES ARE ENTITIES. In addition to that, there is also information about related metaphors and semantic annotation for each language.

Another regularity was the directly related. Frame-to-frame relations[2] are the evidence for such a match. Table 2 is an example of this pattern. The Lexical Units *conduzir.v*, *take.v*, *gedacht.v* are examples of the same metaphor: LIFE IS A JOURNEY.

The frame evoked in English and Brazilian Portuguese is `Bringing`, while in German it is `Cause_motion`. Checking the frame network, there is a direct relation between them, as `Cause_motion` is used by `Bringing`. In this relation, not all Frame Elements of `Cause_motion` occur linguistically in `Bringing`. However, there is a part of its structure presupposed as what is considered a conceptual background.

Table 3 is an example of a situation in which the metaphorical behavior is identical in the three languages through the use of the metaphor TALENT IS AN OBJECT. However, the frame evoked by the Lexical Unit *find.v* in English is `Locating`, while *achar.v* in Brazilian Portuguese and *finden.v* in German is `Becoming_Aware`. Analysing both frames, `Locating` is defined by a Perceiver looking for something, a Sought_entity. And `Becoming_aware`'s definition says that words in this frame have to do with a Cognizer adding some Phenomenon to their model of the world.

Checking the network of frames, we notice that `Locating` uses `Seeking`, which, through the *See_also* relation, links to the `Scrutiny` frame, which, in turn, *uses* `Becoming_aware`. Even though they are indirectly related on the network of FrameNet frames, the annotation divergence seems to be in the choice between the metaphorical source and target domains. `Locating` is a frame related to the source domain, while `Becoming_aware` relates to the metaphorical target domain.

The other pattern in frame comparison was unrelated. In Table 4, the three sentences are an example of the TALENT IS A RESOURCE metaphor. In FrameNet, the lemma *squander.v* is a Lexical Unit in `Expand_resource` and also in `Frugality`. The annotation choice was to insert it in `Expand_resource`, which outlines the use of a resource. In Brazilian Portuguese and in German, the Lexical Units *desperdiçar.v* and *vergeuden.v* evoke `Frugality`, which focuses on how the resource is used.

[2]The frame network is accessed from FrameGrapher on FrameNet. The frame-to-frame relations are Inheritance, Subframe, Perspective_on, Using, Precedes, Inchoative_on, Causative_on, See_also and the Metaphor relation, which was added to the others in the Berkeley team's last systematic update and still lacks empirical validation (Ruppenhofer et al., 2016).

[2]The signal is used when the sentence segmentation is different.

	Portuguese	English	German
IdSentence	820	1087	1348
Sentence	E quando chegam à fase adulta, a maioria das crianças **perdeu** essa capacidade.	And by the time they get to be adults, most kids have **lost** that capacity.	Wenn sie erst erwachsen sind, haben die meisten Kinder diese Fähigkeit **verloren**.
Metaphor behaviour relating to the English text	Metaphor to same metaphor		Metaphor to same metaphor
Metaphor	ATTRIBUTES ARE ENTITIES		
Related metaphor	ATTRIBUTES ARE POSSESSIONS (Grady, 1998; Lakoff, 1999)		
Lexical Unit evoked	*perdeu.v*	*lost.v*	*verloren.v*
Semantic frame	`Losing`	`Losing`	`Losing`
Semantic annotation	[a maioria das crianças $_{\text{Owner}}$] PERDEU [essa capacidade $_{\text{Possession}}$]	[most kids $_{\text{Owner}}$] have LOST [that capacity $_{\text{Possesssion}}$]	[Kinder $_{\text{Owner}}$] diese [Fähigkeit $_{\text{Possesssion}}$] VERLOREN
Match Levels	Total		Total

Table 1: Total semantic frame match level.

	Portuguese	English	German
IdSentence	762	1034	1298
Sentence	Nos interessamos tanto por ela em parte porque é da educação o papel de nos **conduzir** a esse futuro misterioso.	We have a huge vested interest in it, partly because it's education that's meant to **take** us into this future that we can't grasp.	Wir haben ein großes, persönliches Interesse, teilweise Bildung dazu gedacht ist, uns in diese Zukunft zu **bringen**, die wir nicht fassen können.
Metaphor behaviour relating to the English text	Metaphor to same metaphor		Metaphor to same metaphor
Metaphor	LIFE IS A JOURNEY		
Related metaphor	PROGRESSING THROUGH LIFE IS MOVING ALONG A PATH		
Lexical Unit evoked	*conduzir.v*	*take.v*	*bringen.v*
Semantic frame	`Bringing`	`Bringing`	`Cause_motion`
Semantic annotation	[educação Agent] o papel de [nos Theme] CONDUZIR [a esse futuro misterioso Goal]	[education Agent] [that Agent]'s meant to TAKE [us Theme] [into this future that we can't grasp Goal]	[uns Theme] [in diese Zukunft Goal] zu BRINGEN
Match Levels	same		directly related

Table 2: Total semantic frame match level and directly related.

	Portuguese	English	German
IdSentence	772	1043	1307
Sentence	O que vemos ali é uma pessoa de extrema dedicação que **achou** seu talento.	What you have there is a person of extraordinary dedication who **found** a talent.	Sie ist eine Person mit außerordentlicher Hingabe, die ihr Talent **gefunden** hat.
Metaphor behaviour relating to the English text	Metaphor to same metaphor		Metaphor to same metaphor
Metaphor	TALENT IS AN OBJECT		
Related metaphor	A TALENT IS A RESOURCE IDEAS ARE OBJECTS (Lakoff, 1987)		
Lexical Unit evoked	*achou.v*	*found.v*	*gefunden.v*
Semantic frame	`Becoming_aware`	`Locating`	`Becoming_aware`
Semantic annotation	[uma pessoa de extrema dedicação Cognizer] [que Cognizer] ACHOU [seu talento Phenomenon]	[a person of extraordinary dedication Perceiver] [who Perceiver] FOUND [a talent Sought_entity]	[die Cognizer] [ihr Phenomenon][Talent Phenomenon] GEFUNDEN hat
Match Levels	indirectly related		indirecly related

Table 3: Indirectly related semantic frame match level.

	Portuguese	English	German
IdSentence	774	1045	1308
Sentence	E o **desperdiçamos**, implacavelmente.	And we **squander** them, pretty ruthlessly.	(#)[3]und dass wir sie **vergeuden** und zwar ziemlich rücksichtslos.
Metaphor behaviour relating to the English text	Metaphor to same metaphor		Metaphor to same metaphor
Metaphor	TALENT IS A RESOURCE		
Related metaphor	A TALENT IS AN OBJECT		
Lexical Unit evoked	*desperdiçamos.v*	*squander.v*	*vergeuden.v*
Semantic frame	`Frugality`	`Expend_resource`	`Frugality`
Semantic annotation	E [o Resource] DESPERDIÇAMOS, implacavelmente	And [we Agent] SQUANDER [them Resource], [pretty ruthlessly Manner]	[wir Resource_controller] [sie Resource] VERGEUDEN
Match Levels	unrelated		unrelated

Table 4: Unrelated semantic frame matching level (1).

As much as `Frugality` highlights human social behavior, as mentioned in the frame definition, its conceptualization requires the idea of spending or using a resource. Potentially, a frame-to-frame relation connects both of them. However, it has yet to be more deeply studied and attested. In order to achieve this goal, it will be necessary to update the network of frame-to-frame relations.

In Table 5, we can say THE BODY IS A CONTAINER FOR THOUGHTS, VALUES

	Portuguese	English	German
IdSentence	760	1031	1296
Sentence	Porque é uma dessas coisas **arraigadas** nas pessoas, estou certo?	Because it's one of those things that **goes deep** with people, am I right?	Denn es ist eines dieser Themen, die Leute tief **berühren**, wie Religion, Geld und andere Sachen. (#)
Metaphor behaviour relating to the English text	Metaphor to same metaphor		Metaphor to same metaphor
Metaphor	THE BODY IS A CONTAINER FOR THOUGHTS, VALUES, PRINCIPLES		
Related metaphor	THE BODY IS A CONTAINER FOR EMOTIONS (Lakoff, 1987)		
Lexical Unit evoked	*arraigada.a*	*goes.v / deep.a*	*berühren.v*
Semantic frame	`Presence`	`Motion` / `Measurable_attributes`	`Stimulus_focus`
Semantic annotation	[coisas $_{\text{Entity}}$] ARRAIGADAS [nas pessoas $_{\text{Location}}$]	[those things $_{\text{Theme}}$] [that $_{\text{Theme}}$] GOES [deep $_{\text{Goal}}$]	[Themen, die $_{\text{Stimulus}}$] [Leute $_{\text{Experiencer}}$] [tief $_{\text{Degree}}$] BERÜHREN, [wie Religion, Geld und andere Sachen $_{\text{Comparison\_set}}$]
Match Levels	unrelated		unrelated

Table 5: Unrelated semantic frame matching level (2).

AND PRINCIPLES is the general metaphor in the three languages. However, there are differences in the framing of each one. In English, *go deep* evoke `Motion` and `Measurable_attributes`, while, in German, *berühren.v* (which can be literally translated into English as *touch.v*) evokes `Stimulus_focus`.

In both cases, the metaphor THOUGHTS, VALUES AND PRINCIPLES ARE OBJECTS is used. Unlike previous uses where the metaphor is related to dynamic events, in Brazilian Portuguese, the lexical unit is *arraigado.a* (which can be literally translated into English as *rooted.a*). The lemma was annotated in the `Presence` frame. A specific metaphor that licenses this use is THOUGHTS, VALUES, PRINCIPLES ARE PLANTS. Through this metaphor, just as a plant takes root and becomes fixed in the ground, a thought, value and principle can also become established in a person.

On preliminary analysis, the unrelated pattern includes different situations. Cases in which some relation may exist, but is not in the database, are included here. A possible explanation is the dynamic character of the database updating. Other possible reasons for such a pattern may lie in the perspectives assumed in the face of the linguistic framework of a given situation, as well as features of idiomaticity and typological specificities of languages.

6 Summary

This paper compared metaphors and frames in the FrameNet multilingual annotation task. The result of this process is a set of analyses that indicate paths to a descriptive method of extracting semantic information from FrameNet database. Future work may validate the taxonomy presented in a larger sample of semantically annotated parallel corpus and expand the analysis to other languages. Beyond that, including computational works on multilingual approaches to frame semantics and metaphors can contribute to a method to automatically parameterize these data.

Acknowledgements

The work reported in this article has received financial support from the CAPES PROBAL (grant 88887.387875/2019-00). The author would like to thank Dr. Oliver Czulo who supervised the postdoctoral fellowship at the University of Leipzig in 2020, and Dr. Tiago Torrent who contributed to this research as well. I extend the thanks to Dr. Helen de Andrade, the anonymous reviewers and the editors for their important comments and suggestions to improve the final version of the paper. Finally, I am also thankful for all the researchers who put effort to the shared annotation task and provide their data for studies like this present one.

References

Collin F Baker and Arthur Lorenzi. 2020. Exploring crosslinguisitic framenet alignment. In *LREC 2020 WORKSHOP*, page 77.

Oliver Czulo. 2017. Aspects of a primacy of frame model of translation. *Empirical modelling of translation and interpreting*, 7:465.

Ellen K Dodge, Jisup Hong, and Elise Stickles. 2015. Metanet: Deep semantic automatic metaphor analysis. In *Proceedings of the Third Workshop on Metaphor in NLP*, pages 40–49.

Gilles Fauconnier. 1994. *Mental spaces: Aspects of meaning construction in natural language*. Cambridge University Press.

Charles J Fillmore. 1982. Frame semantics. in.: The linguistic society of korea. linguistics in the morning calm. *Korea: Hanshin Publishing Company*.

Charles J Fillmore. 1985. Frames and the semantics of understanding. *Quaderni di semantica*, 6(2):222–254.

George Lakoff. 2008. *Women, fire, and dangerous things: What categories reveal about the mind*. University of Chicago press.

George Lakoff. 2012. Explaining embodied cognition results. *Topics in cognitive science*, 4(4):773–785.

George Lakoff and Mark Johnson. 1999. *Philosophy in the Flesh: the Embodied Mind & its Challenge to Western Thought*. Basic Books.

George Lakoff and Mark Johnson. 2008. *Metaphors we live by*. University of Chicago press.

George Lakoff and Mark Turner. 2009. *More than cool reason: A field guide to poetic metaphor*. University of Chicago press.

Ronald W Langacker. 1987. *Foundations of cognitive grammar: Theoretical prerequisites*, volume 1. Stanford university press.

Peter Newmark. 1981. *Approaches to translation (Language Teaching methodology senes)*. Oxford: Pergamon Press.

Ken Robinson. Do schools kill creativity? [online]. 2016.

Josef Ruppenhofer, Michael Ellsworth, Myriam Schwarzer-Petruck, Christopher R Johnson, and Jan Scheffczyk. 2016. Framenet ii: Extended theory and practice. Technical report, International Computer Science Institute.

Christina Schäffner. 2004. Metaphor and translation: some implications of a cognitive approach. *Journal of pragmatics*, 36(7):1253–1269.

Christina Schäffner. 2016. Rojo, a., & ibarretxe-antuñano, i.(eds.).(2013). cognitive linguistics and translation: Advances in some theoretical models and applications. *Review of Cognitive Linguistics*, 14(2):523–531.

Tiago Timponi Torrent, Michael Ellsworth, Collin Baker, and EE Matos. 2018. The multilingual framenet shared annotation task: a preliminary report. In *Proceedings of the Eleventh International Conference on Language Resources and Evaluation*, pages 62–68.

Tiago Timponi Torrent, Ely Edison da Silva Matos, Frederico Belcavello, Marcelo Viridiano, Maucha Andrade Gamonal, Alexandre Diniz da Costa, and Mateus Coutinho Marim. 2022. Representing Context in FrameNet: A Multidimensional, Multimodal Approach. *Frontiers in Psychology*, 13.

Gideon Toury. 1995. *Descriptive translation studies and beyond*, volume 4. J. Benjamins Amsterdam.

Language Resource References

FrameNet:
`https://framenet.icsi.berkeley.edu`

MetaNet:
`https://metaphor.icsi.berkeley.edu`

Multilingual FrameNet iniciative:
`https://www.globalframenet.org`

Multi-Sense Language Modelling

Andrea Lekkas[1,2,3], **Peter Schneider-Kamp**[1], **Isabelle Augenstein**[3]
[1]Dept. of Mathematics and Computer Science, University of Southern Denmark
[2]Ordbogen A/S, Odense, Denmark
[3]Dept. of Computer Science, University of Copenhagen, Denmark
gzt740@alumni.ku.dk, petersk@imada.sdu.dk, augenstein@di.ku.dk

Abstract

The effectiveness of a language model is influenced by its token representations, which must encode contextual information and handle the same word form having a plurality of meanings (polysemy). Currently, none of the common language modelling architectures explicitly model polysemy. We propose a language model which not only predicts the next word, but also its sense in context. We argue that this higher prediction granularity may be useful for end tasks such as assistive writing, and allow for more a precise linking of language models with knowledge bases. We find that multi-sense language modelling requires architectures that go beyond standard language models, and here propose a localized prediction framework that decomposes the task into a word followed by a sense prediction task. To aid sense prediction, we utilise a Graph Attention Network, which encodes definitions and example uses of word senses. Overall, we find that multi-sense language modelling is a highly challenging task, and suggest that future work focus on the creation of more annotated training datasets.

1 Introduction

Any variant of language model, whether standard left-to-right, masked (Devlin et al., 2019) or bidirectional (Peters et al., 2018) has to address the problem of *polysemy*: the same word form having multiple meanings, as seen in Tab. 1. The meaning of a particular occurrence depends on the context, and all modern language modelling architectures from simple RNNs (Mikolov et al., 2010) to Transformers (Vaswani et al., 2017) use context-based representations. However, *token* representations in language models are not explicitly disambiguated. Single-prototype embeddings, i.e., traditional word vectors, have a 1-to-1 correspondence with word forms. Contextual embeddings change depending on the tokens in their context window, and are employed in recent models like ULMFit (Howard and Ruder, 2018), ELMo (Peters et al., 2018), and all Transformer architectures. However, even for contextual embeddings, polysemy is handled in an implicit, non-discrete way: the sense that a word assumes in a particular occurrence is unspecified.

Sentence	Meaning
*"John sat on the **bank** of the river and watched the currents"*	**bank.n.01**: *sloping land, especially the slope beside a body of water*
*"Jane went to the **bank** to discuss the mortgage"*	**bank.n.02**: *a financial institution that accepts deposits and channels the money into lending activities*

Table 1: Example of polysemy. Senses taken from WordNet 3.0

Here, we propose the task of multi-sense language modelling, consisting of not only word, but also sense prediction. We conjecture that multi-sense language modelling would:

1) improve the precision of linking a language model to a knowledge base, as done in Logan et al. (2019), to help generate factually correct language. For instance: *"The explorers descended in the cave and encountered a bat"* refers to the entity *'bat (animal)'* and not to *'bat (baseball implement)'*.
2) be useful in applications such as assistive writing (Chen et al., 2012), where it is desirable to display more information about a word to a user, such as its sense, definition or usage.

Another potential use would be to explore if such dictionary information could improve standard language modelling, and reduce the number of training data needed, relevant for e.g. language modelling for low-resource languages.

Consequently, our **research objectives** are to:

- model next-sense prediction as a task alongside standard next-word prediction in lan-

Proceedings of the Workshop on Dimensions of Meaning: Distributional and Curated Semantics (DistCurate 2022), pages 8 - 18
July 14, 2022

guage modelling, and examine the performance of different model architectures.
- encode sense background knowledge from sense definitions and examples, and examine how it can aid sense prediction.

As a sense inventory, we use WordNet 3.0 (Miller, 1995). The sense background knowledge is encoded in a dictionary graph, as shown in Fig. 3. When reading a word w, the model can rely on an additional input signal: the state of the node that represents w in the dictionary graph (the *"global node"*). Node vectors in the graph are updated by a Graph Attention Network (Veličković et al., 2018).

Findings: We find that sense prediction is a significantly more difficult task than standard word prediction. A way to tackle it is to use a localized prediction framework, where the next sense depends on the prediction of the next word. The most successful model we identified for this uses a hard cut-off for the number of words considered (SelectK, see § 4.4). The additional input signal from the dictionary graph provides only marginal improvements. For future work, we argue that multi-sense language modelling would benefit from larger sense-labelled datasets, possibly aided by WordNet super-sense information (categories like food, artifact, person, etc.) to deal with the excessively fine granularity of WordNet senses.

2 Related Work

We here briefly discuss relevant works that disambiguate between word senses to address polysemy. They can be grouped in three categories: 1) multi-prototype embeddings not connected to a knowledge base; 2) supervised multi-sense embeddings based on a text corpus that only utilise a KB tangentially as the sense inventory; and 3) models that rely more substantially on features from a KB, like glosses or semantic relations.

Multi-prototype embeddings Huang et al. (2012) learn multi-prototype vectors by clustering word context representations. Single-prototype embeddings are determined by 2 FF-NNs with a margin objective on predicting the next word, a quasi-language modelling setting even if it utilises both the preceding and subsequent context. Multi-sense skip-gram (Neelakantan et al., 2014) also defines senses as cluster centroids, measuring the cosine distance of the surrounding context of ± 5 words. Li and Jurafsky (2015) use Chinese Restaurant Processes to decide whether to create a new cluster, and also investigate the usefulness of multi-sense embeddings in downstream tasks such as Semantic Relatedness and PoS tagging. (Chronis and Erk, 2020) create multi-prototype embeddings from BERT, to address word similarity and relatedness tasks. Every word w is associated with a set of occurrence embeddings, which are computed by averaging sub-word WordPiece tokens. Senses are obtained by applying K-means clustering to the set. Each BERT layer contributes a different set, thus the authors found that middle layers are more relevant to the similarity task, the final layers to relatedness.

Supervised multi-sense embeddings Other models rely on sense-label supervision in a text corpus. *context2vec* (Melamud et al., 2016) builds contextual word embeddings by applying a biLSTM on text, and provides the option to create sense embeddings by using a sense-labelled corpus. Raganato et al. (2017) frames WSD as a sequence learning problem, with the aim of finding sense labels for an input sequence. The training corpus is the same used in our work, SemCor (Miller et al., 1993), and the core architecture is a biLSTM that reads both the preceding and subsequent context. A biLSTM is also employed in LSTMEmbed (Iacobacci and Navigli, 2019), that obtains a sense-labelled training corpus by applying the BabelFly (Moro et al., 2014) sense tagger on the English Wikipedia and other texts.

Recently, SenseBERT (Levine et al., 2020) uses a BERT transformer encoder with two output mappings, one for the MLM-masked words, another for their WordNet supersenses. SenseBERT relies on soft labeling, associating a masked word w with any one of its supersenses $S(w)$. Using very large text corpora is expected to reinforce the correct supersense labels.

KB-based methods Sense representations can leverage WordNet glosses, as done in Chen et al. (2014) after single-prototype vectors are trained with a skip-gram model. Likewise, pre-trained word embeddings are the starting point for AutoExtend (Rothe and Schütze, 2015), an autoencoder architecture where word embeddings constitute the input and the target whereas the embeddings for WordNet synsets are the intermediate encoded representation. Kumar et al. (2019) use a BiLSTM and self-attention to get contextual embeddings, then disambiguate them via a dot product with sense embeddings based on WordNet definitions.

In the last couple of years, efforts have been made to enable BERT to disambiguate between senses. GlossBERT (Huang et al., 2019) takes in context-gloss pairs as input: for every target word in a context sentence, N=4 WordNet glosses are found; a classification layer determines which lemma the word assumes. SenseEmBERT (Scarlini et al., 2020a) relies on the BabelNet mapping between WordNet and Wikipedia pages to collect relevant text for synsets, and computes the sense embeddings as a rank-weighted average of relevant synsets. Scarlini et al. (2020b) applies the same pipeline of of context extraction, synset embeddings and sense embeddings construction: it computes lemma representations via BERT, and uses UKB (Agirre et al., 2014) to create a set of contexts for the synset. Sense embeddings are obtained by concatenating the BERT representation of the sense contexts found in SemCor and the sense definition in WordNet.

Our model We use pre-trained embeddings and WordNet glosses and relations to create a dictionary graph. The vectors of sense nodes can be viewed as sense embeddings; we are not primarily interested in their quality since our objective is not "classic" Word Sense Disambiguation or relatedness, but rather multi-sense language modelling. Future work may rely on them to improve sense disambiguation: for model variants that have to choose the correct sense among a limited number of candidates, there is an opening for the application of more complex multi-sense models than the ones explored here.

3 Multi-Sense Language Model

3.1 Architecture

A language modelling task decomposes the probability of predicting an entire text of length N as the product of each word prediction, where the probability of the next word $p(w_i)$ is influenced by the preceding context $[w_1, ..., w_{i-1}]$:

$$p(w_1, ..., w_N) = \prod_{i=1}^{N} p(w_i | w_1, ..., w_{i-1}) \quad (1)$$

Our model aims to carry out two language modelling tasks:

1. Standard language modelling: next-token prediction at the granularity of words.

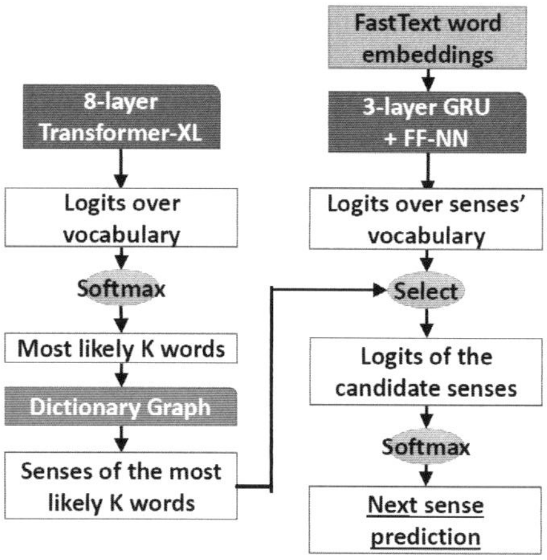

Figure 1: Overview of the SelectK version of the model that uses a Transformer-XL for the StandardLM task.

2. Sense prediction: next-token prediction at the granularity of WordNet senses.

Therefore, the objective is to produce, at each position in the corpus, two probability distributions: one over the vocabulary of words and one over the vocabulary of senses.

The architecture for the standard language modelling task is either a 3-layer GRU followed by a FF-NN, or an 8-layer Transformer-XL (Dai et al., 2019). Both are pre-trained on WikiText-2 (Merity et al., 2016), and then the whole model is trained on the SemCor sense-labeled corpus (Miller et al., 1993). Some experiments use a Gold LM, that always predicts the correct next word, with the aim to examine the effectiveness of sense architectures independently from the accuracy of the standard language modelling task.

The input signal for the sense prediction architecture consists of FastText pre-trained embeddings (Bojanowski et al., 2017), possibly augmented by concatenating the graph node corresponding to the current word, as shown in Fig. 2. Moreover, a model variant that uses localized prediction also relies on the K most likely next words predicted by the standard language modeling task: $w^1_{t+1}, \ldots, w^K_{t+1}$. An example of a localized prediction variant, SelectK (§ 4.4), is shown in Fig. 1.

The correctness of the prediction is evaluated using two measures: perplexity and accuracy.

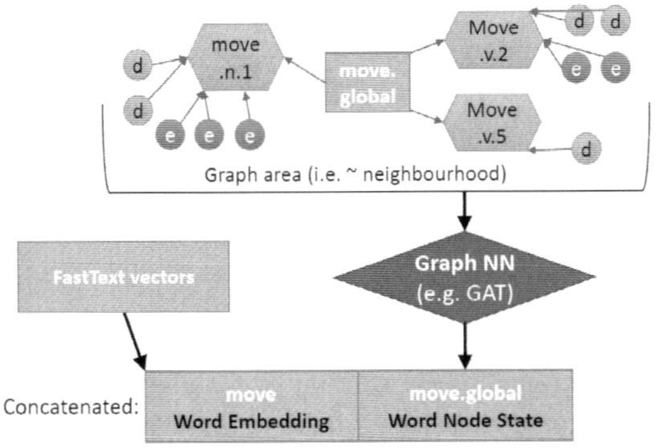

Figure 2: The input signals for graph-informed sense prediction: the standard word embedding and the vector of the global node from the dictionary graph

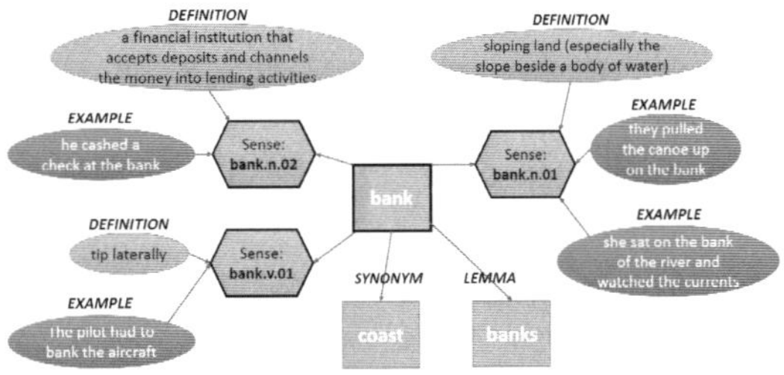

Figure 3: Part of the dictionary graph for the word *"bank"*. **Global** nodes are in yellow, sense nodes in green, definitions and examples in light blue and blue.

3.2 Dictionary Graph

The purpose of the dictionary graph is to provide a way for the input signal to be informed by the sense embeddings and possibly to leverage WordNet's graph structure and glosses.

First, we read in a text corpus and create the **vocabulary**. We register the lemmatised version of inflected forms, later connected to their parent forms. Then, for each sense of a word, we retrieve the text of its **definitions** and **examples**, and register the connections with synonyms and antonyms. In WordNet, senses are specified as the synsets of a word; example: w='bank' $\rightarrow$ [('bank.n.01'), ... ,('bank.v.07)] .

The next step is to compute the **sentence embeddings** for definitions and examples. It is possible to use any one of several methods, ranging from LSTMs to BERT's last layer. For the sake of computation speed, sentence embeddings are obtained as the average of the FastText word vectors.

Finally, the nodes are initialised and stored in a graph with their edges. The graph object is created with PyTorch-Geometric (Fey and Lenssen, 2019); before training, it holds the initial values for the sense and global nodes. The **sense node** starts as the average of the embeddings of definitions and examples for that sense. As shown in Fig. 3, every sense node is directly connected to its definition and example nodes. Those determine its starting position; ideally, during training, it would be moved towards the more significant glosses. The **global node** is initialised as the FastText embedding for w.

3.3 Graph Attention Network

We employ a Graph Attention Network (Veličković et al., 2018) to update the nodes of the dictionary graph. Unlike Graph Convolutional Networks (Kipf and Welling, 2017), a GAT does not require a fixed graph structure, and can operate on different local graph-batches like the neighbourhood of the current word. Unlike other methods like graphSAGE (Hamilton et al., 2017) it can handle a variable number of neighbours. The underlying idea of GATs is to compute the representation vector h_i of node i based on its neighbouring nodes $j \in N(i)$, which have different attention weights (i.e. importance).

We here describe how a GAT obtains the new state h_i^{t+1} of node i in a graph with m nodes. First, a linear transformation $\mathbf{W}$ is applied over all the nodes: $\mathbf{W}h_1^t, ..., \mathbf{W}h_m^t$. Then, for each node j in the neighbourhood of i, we compute the non-normalised attention coefficient e_{ij}, via a 1-layer FF-NN with LeakyReLU activation:

$$e_{ij} = LeakyReLU(\mathbf{A}^T[\mathbf{W}h_i, \mathbf{W}h_j]) \quad (2)$$

The normalised attention coefficients α_{ij} are obtained by applying a softmax over the neighbourhood of i, $N(i)$. Finally, the new state of node i is given by applying a non-linear function ρ to the weighted sum of the neighbours' states:

$$h_i^{t+1} = \rho\left(\sum_{j \in N(i)} \alpha_{ij}\mathbf{W}h_j^t\right) \quad (3)$$

Veličković et al. (2018) report that using multiple attention heads, by averaging or by concatenating, is beneficial to stabilise the model; we therefore use 2 concatenated heads.

	Words		Senses	
	PPL	Accuracy	PPL	Accuracy
2 GRUs	160.94	0.209	**562.46**	0.053
2 Transformer-XL	128.99	0.241	**186.72**	0.217

Table 2: Baselines: two separate GRUs, and two separate Transformer-XL models. Results of word and sense prediction on SemCor's test set.

3.4 Dealing with Missing Sense Labels

Even in a sense-labelled corpus, some of the words, such as stopwords, will not have a sense label. This may cause a problem because in some versions of the model (§ 4.4, §4.5, §4.6), the GRU used for the sense prediction task should be able to read the input text as an uninterrupted sequence of sense tokens. In particular, two types of words may have no sense specification:

1. **stopwords**: *'for'*, *'and'*, *'of'*, etc.
2. **inflected forms**: *'is'*, *'said'*, *'sports'*, etc.

In order to provide stopwords with a sense label, we add a **dummySense** (e.g. 'for.dummySense.01') for all the words without a sense. The corresponding graph node is initialised as the single-prototype FastText vector for that word. Inflected forms are **lemmatised** using NLTK's WordNetLemmatizer, to be able to read and predict the senses of their parent form ('is'→'be', 'said'→'say' etc.)

4 Architectures for Sense Prediction

As previously described, multi-sense language modelling consists of the two tasks of standard language modelling and sense prediction, and we aim to output two probability distributions - for the next word token and the next sense token, respectively.

4.1 GRU

A 3-layer GRU followed by a 1-layer FF-NN. There are no shared layers with the standard language modelling architecture, whether that be a Transformer-XL or another GRU. In the latter case, the 2 GRUs share the input signal, i.e. the FastText embeddings, possibly concatenated with the word node embedding.

4.2 Transformer-XL

Transformer-XL is a left-to-right transformer encoder that operates at word-level granularity. It was chosen instead of a BERT model because the latter uses masked language modelling with WordPiece sub-word tokenisation, and thus cannot easily be adapted to a setting where the the next word and its sense should be predicted. In our setting, the Transformer-XL learns word embeddings from scratch and does not rely on the FastText vectors.

4.3 Most Frequent Sense

This heuristic baseline chooses the *most frequent sense* found in the training set for the most likely word predicted by the standard language model.

4.4 SelectK

SelectK is a localized prediction approach: first the next word is predicted, giving a set of K most likely candidates; then, the next sense is chosen among them.

As the text is read, for every location t, the standard language model outputs a probability distribution over the vocabulary, where the most likely K words are $w_1, ..., w_K$. Every word w_i has a set of senses: $S(w_i) = \{s_{i1}, ..., s_{iN}\}$. The next sense at location t is chosen among the senses of the K most likely words:

$$s(t) \in \bigcup_{i=1}^{K} S(w_i) \tag{4}$$

A softmax function is applied over the logits of the selected senses, while all other senses are assigned a probability $\epsilon = 10^{-8}$. The senses' logits are computed by a dedicated GRU, as described in Figure 1. Alternatively, a Transformer-XL could be used, but it would presumably require more training data than the SemCor corpus alone.

K is a hyperparameter; K=1 means that the model always chooses among the senses of the most likely word. In general, the sense prediction performance depends on the performance of the standard language model: if all the K most likely globals are incorrect, the correct sense cannot be retrieved. We verify what happens for K={1,5,10}.

4.5 Sense Context Similarity

Another localized prediction method is to select the senses of the most likely K globals as candidates; then, rank them based on the cosine similarity between the local context and each sense's average context. Since language modelling is performed from left to right, the context is based only on the **preceding** tokens $[w_{t-1}, ..., w_{t-c}]$ without considering subsequent tokens.

Methods & Parameters				SemCor test			SensEval dataset		
Senses architecture	Standard LM	K	context	Senses ACC	Polysem ACC	Globals ACC	Senses ACC	Polysem ACC	Globals ACC
MFS	**Gold**	**-**	**-**	0.83	0.62	1	**0.82**	**0.62**	**1**
SelectK	**Gold**	**1**	**-**	**0.90**	**0.80**	**1**	0.82	0.61	1
SenseContext	Gold	1	average	0.73	0.45	1	0.71	0.41	1
Self-attention	Gold	1	average	0.52	0.24	1	0.51	0.25	1
MFS	**TXL**	**-**	**-**	**0.24**	**0.02**	**0.24**	**0.23**	**0.04**	**0.23**
TXL	TXL	-		0.22	0.02	0.24	0.20	0.03	0.22
SelectK	**TXL**	**1**	**-**	**0.24**	**0.02**	**0.24**	0.23	0.03	0.23
SelectK	TXL	5	-	0.13	0.01	0.24	0.11	0.01	0.22
SenseContext	TXL	1	average	0.22	0.01	0.24	0.22	0.02	0.23
SenseContext	TXL	1	GRU	0.19	0.00	0.24	0.17	0.00	0.23
SenseContext	TXL	5	average	0.13	0.01	0.24	0.12	0.01	0.22
Self-attention	TXL	1	average	0.17	0.01	0.24	0.16	0.02	0.23
Self-attention	TXL	5	average	0.08	0.01	0.24	0.07	0.01	0.23
MFS	**GRU**	-	-	**0.22**	**0.02**	**0.22**	**0.22**	**0.03**	**0.22**
SelectK	**GRU**	**1**	-	**0.22**	**0.02**	**0.22**	0.22	0.02	0.21
SenseContext	GRU	1	average	0.20	0.00	0.22	0.20	0.00	0.22
Self-attention	GRU	1	average	0.16	0.01	0.22	0.16	0.02	0.22
SelectK	GRU	5	-	0.13	0.00	0.21	0.12	0.01	0.21
GRU	GRU	-	-	0.05	0.00	0.21	0.06	0.00	0.21

Table 3: The most relevant results of each method. Task: sense prediction. Datasets: SemCor's test split (10%) and the aggregated SensEval dataset by Raganato et al. (2017) .

For each occurrence $s_1, ...s_N$ of a sense s, the occurrence context $OC(s_i)$ is computed as the average of the word embeddings of the preceding c tokens. Afterwards, the Sense Context $SC(s)$ is computed as the average of the occurrences' contexts:

$$OC(s_i) = avg(w_{t-1}, ..., w_{t-c}) \quad (5)$$
$$SC(s) = avg(OC(s_1), ..., OC(s_N))$$

We also experiment with obtaining a representation of the local context with a 3-layer GRU.

4.6 Self-Attention Coefficients

Another way to choose among the candidate senses of the most likely K globals is to use the softmax scores from the self-attention mechanism. Every sense s has an average context $SC(s)$ it appears in, as seen in Eq. 5. The contexts of the candidate senses are collected in the matrix C. Then, a probability distribution over the senses is obtained by computing the self-attention coefficients:

$$softmax\left(\frac{Q \cdot C}{\sqrt{d_k}}\right) \quad (6)$$

All the rows of the query matrix Q are representations of the current context. As previously, the local context can be constructed as a simple average of the last c word embeddings, or as the output of a 3-layer GRU. The sense contexts in C take up the role of keys in the formula of self-attention scores.

5 Evaluation

5.1 Dataset and Graph Settings

To train a multi-sense language model, we need a sense-labelled text corpus. We use **SemCor** (Miller et al., 1993), a subset of the Brown Corpus labelled with senses from WordNet 3.0. Training, validation and test sets are obtained with a 80/10/10 split. Since standard LM architectures are pre-trained on WikiText-2, the vocabulary is obtained from WikiText-2 and SemCor's training split (the latter with min. frequency=2). The dictionary graph has ≈169K nodes and ≈254K edges. Consequently, due to memory constraints, we apply mini-batching on the graph: for each input instance the Graph Attention Network operates on a local **graph area**. The graph area is defined expanding outwards from the global node of the current word w. In our experiments, a graph area contains a maximum of 32 nodes and extends for only 1 hop, thus coinciding with the **neighbourhood** of the current word's global node.

5.2 Model variants

In Tab. 3 and 4, we show the results of sense predictions with different architectures and parameters.

Methods & Parameters				SemCor test set				SensEval dataset			
Senses architecture	Standard LM	K	context	Senses ACC	DG Δ	PPL	DG Δ	Senses ACC	DG Δ	PPL	DG Δ
SelectK	TXL	1	-	**0.24**	**0**	119.6	-0.1	**0.23**	**0**	172.54	-1.34
SelectK	TXL	5	-	0.13	0	121.0	0	0.11	0	162.2	+0.3
SenseContext	TXL	1	average	0.22	0	118.9	+1.2	0.22	0	172.4	+8.0
SenseContext	TXL	5	average	0.13	0	135.8	+0.5	0.12	0	213.2	+2.1
Self-attention	TXL	1	average	0.17	+0.01	**111.9**	**-7.3**	0.17	+0.01	**155.7**	**-16.8**
Self-attention	TXL	5	average	0.08	0	120.6	+1.3	0.08	+0.01	161.7	-4.6
GRU	GRU	-	-	0.05	0	148.5	-12.5	0.06	0	157.7	-5.8
SelectK	GRU	1	-	**0.22**	**0**	131.7	-1.6	**0.22**	**+0.01**	**153.1**	**-3.6**
SelectK	GRU	5	-	0.13	-0.01	148.5	-12.1	0.11	-0.01	157.7	-5.6
SenseContext	GRU	1	average	0.21	+0.01	129.5	-3.7	0.21	+0.01	153.3	-3.4
SenseContext	GRU	5	average	0.09	+0.01	141.9	-18.5	0.10	+0.02	182.0	+18.7
Self-attention	GRU	1	average	0.16	+0.01	**128.1**	**-10.3**	0.17	+0.01	154.0	-2.7
Self-attention	GRU	5	average	0.08	0.00	139.8	-20.7	0.07	0.00	154.8	-9.5

Table 4: Word and sense prediction. **DG Δ** = change caused by including the input from the dictionary graph.

Sense prediction architectures:

- **GRU**: a 3-layer GRU followed by FF-NN
- **TXL**: an 8-layer Transformer-XL
- **MFS**: Given the most likely word predicted by the standard LM, choose its most frequent sense in the training set (see § 4.3).
- **SelectK**: A GRU computes the logits over the senses' vocabulary, but the softmax is restricted to the *candidate senses* only: the senses of the most likely K words from the standard LM task (see § 4.4).
- **SenseContext**: Choosing among the candidate senses based on the cosine similarity of local context and average sense context (see § 4.5).
- **Self-Attention**: Choosing among the candidate senses by computing the self-attention coefficients of local context and average sense contexts (see § 4.6).

Tab. 2 compares the simplest methods available: using either two GRUs or two Transformer-XL for the two tasks of word and sense prediction. While the Transformer-XL performs better than the GRU, these are not the best results: further experiments shed light on the sense prediction model variants.

5.3 Results

Overall Results In Tab. 3, we compare the results of the different methods for sense prediction. We report the **Senses ACC**, accuracy on the senses of all words, and the **Polysem ACC**: accuracy on the senses of polysemous words; words that have more than 1 sense are an important part of any multi-sense task, and are expected to be more difficult. We include the accuracy for standard word prediction, **Globals ACC**, because in our structural prediction setting, if the next word is incorrect, none of its candidate senses will be correct.

We evaluate sense prediction using accuracy instead of perplexity, since in localized prediction methods, perplexity values are non-significant ($\approx 10^8$), due to having assigned $\epsilon = 10^{-8}$ as the probability of the non-candidate senses.

Experiments with the Gold standard language model allow us to examine the methods' capability to discriminate between senses if the next word is always predicted correctly.

The only strong accuracy values (>50%) are obtained by methods using the Gold standard language model, due to the dependency on word prediction. On the SemCor test split, the best-performing method is SelectK with K=1, which is extremely reliant on the correctness of the word prediction task. On the aggregated dataset of the SemEval-SensEval tasks (Raganato et al., 2017), picking the most frequent sense is a difficult baseline to beat (with Senses ACC=0.82 and Polysem ACC=0.62), but SelectK with K=1 is extremely close.

We found that increasing K to 5 or 10 leads to a worse performance, due to the increased difficulty of choosing among a greater number of candidate senses. Using a Transformer-XL as Standard LM gives a small improvement over the GRU. The context representation made from the average of the last 20 tokens is better than the one created by a dedicated GRU trained on SemCor.

Predictions for Polysemous Words All non-Gold model variants have very low accuracy when it comes to predicting the right sense of a polysemous word. This is due to polysemous words being the most difficult ones to predict correctly in the standard language modelling task, and therefore in a localized prediction framework.

This is corroborated by observing the word predictions of the Transformer-XL standard language model. The first batch of the SensEval dataset has 130 correct predictions out of 512 samples, where the correct words and their frequency are: {'of':10, 'the':27, '<unk>':26, 'and':3, ',':16, '.':10, '<eos>':16, 'is':5, 'to':6, 'have':1, 'a':2, 'be':2, '"':1, 'one':1, ''s':1, 'ago':1, 'with':1, 'are':1}

Even if the Perplexity values are reasonable (with a Transformer-XL, $\sim$ 170 on SensEval and $\sim$ 120 on the SemCor test set), most polysemous words are not going to be predicted correctly. Consequently, a way to improve multi-sense language modelling would be to utilise a better-performing model for the standard language modelling task. Sense prediction itself could likely be improved by training on a larger sense-annotated training dataset than SemCor.

Inclusion of Dictionary Graph Input The input signal can be the concatenation of the FastText vector and graph node for the word *w*, as shown in Fig. 2. This is used in two occurrences: first, by the GRU-based standard LM; secondly, by the sense architectures in the SelectK, SenseContext and Self-Attention variants. We aim to investigate whether the dictionary graph input improves the GRU-based word prediction and the local sense prediction.

Tab. 4 shows that the impact of the graph input signal on sense prediction is negligible, while giving a slight boost to the Self-Attention and SenseContext methods. Moreover, it often produces a small perplexity improvement for the GRU standard language model.

Future work may research how to make the graph input more helpful for word and sense prediction, by examining the use of: different Graph Neural Networks, different parameters, or different ways of encoding dictionary definitions and examples

6 Discussion

As shown in Tab. 2, next-token prediction at the granularity of senses is a more difficult task than standard language modelling, due to operating on a larger vocabulary with a more extensive low-frequency long tail. To try to overcome this obstacle, we proposed a localized prediction framework, finding that it is extremely reliant on the correctness of standard language model prediction.

An argument can be made for developing better sense discrimination models that can work with a higher number of candidate senses, for instance those deriving from K=5 instead of K=1. With this in mind, we observe that there are relatively few sense-labelled datasets. The datasets organised in UFSAC format (Vial et al., 2018) altogether contain 44.6M words, of which only 2.0M are annotated; in SemCor, 29.4% of the tokens have a sense label. These datasets are available for English only, thus, studying the benefit of using dictionary resources for low-resource languages cannot currently be pursued until such corpora are created.

As seen in Tab. 3, the best results we managed to achieve are obtained by choosing among the senses of the most likely word. If a sense prediction method managed to reliably choose among a higher number of candidate senses, it would make the sense prediction task less dependant on achieving a good performance for the standard language modelling task. The question of what such a method could be remains open. It may be solved by investigating different WSD methods, and possibly by different ways of encoding the dictionary graph. Moreover, one could expect that next-token prediction, both at the word and the sense level, would benefit from a more accurate standard language modelling architecture, possibly pre-trained on a larger corpus than WikiText-2. However, WikiText-2 was chosen to avoid overwhelming SemCor's vocabulary, so this brings us back to the necessity of a larger sense-labelled corpus, that would also allow one to use an architecture different than a GRU to obtain logits over the senses.

Including the input signal from the dictionary graph only results in marginal improvements on the sense prediction task. It should be noted that the quality of the input signal is limited by the quality of the sentence encodings for the WordNet glosses, used to initialise the graph nodes. Sentence representations different from averaging FastText embeddings may achieve better results. Moreover, tuning the graph signal is surely possible, while outside of the scope of this first study of multi-sense language modelling: one could experiment with changing the size of the graph area, the number of hops and the variant of Graph Neural Network used.

7 Conclusions

This work constitutes the first study of multi-sense language modelling. We experiment with a localized prediction approach that predicts a word followed by a word sense; as well as learning sense representations from both its sentential context, and a sense dictionary, encoding the latter using a Graph Neural Network.

The experimental results highlight the difficulty of such a novel task. Some could regard word senses as not fit to be discretized in a language modeling task, and best represented by flexible contextual embeddings like those of transformers. We believe that specifying discrete senses in language modeling could still be improved investigating three directions:

1) Training a model on larger sense-labelled resources;
2) Using different tools to build the models; for instance, creating word embeddings from BERT by averaging the WordPiece-encoded tokens, or applying other Word-sense Disambiguation methods;
3) Predicting WordNet supersenses: higher level categories such as food, artifact, person; this would avoid relying on the fine granularity of WordNet senses, solving a relatively simpler task that would still provide useful distinctions.

Future work on this task may view it as a test bed for researching Word Sense Disambiguation, as a way of improving the precision of linking a language model to a knowledge base, or for applications such as assistive writing.

Acknowledgments

This work is partly funded by Innovation Fund Denmark under grant agreement number 8053-00187B.

References

Eneko Agirre, Oier López de Lacalle, and Aitor Soroa. 2014. Random walks for knowledge-based word sense disambiguation. *Computational Linguistics*, 40(1):57–84.

Piotr Bojanowski, Edouard Grave, Armand Joulin, and Tomas Mikolov. 2017. Enriching word vectors with subword information. *Transactions of the Association for Computational Linguistics*, 5:135–146.

Mei-Hua Chen, Shih-Ting Huang, Hung-Ting Hsieh, Ting-Hui Kao, and Jason S. Chang. 2012. FLOW: A first-language-oriented writing assistant system. In *Proceedings of the ACL 2012 System Demonstrations*, pages 157–162, Jeju Island, Korea. Association for Computational Linguistics.

Xinxiong Chen, Zhiyuan Liu, and Maosong Sun. 2014. A unified model for word sense representation and disambiguation. In *Proceedings of the 2014 Conference on Empirical Methods in Natural Language Processing (EMNLP)*, pages 1025–1035, Doha, Qatar. Association for Computational Linguistics.

Gabriella Chronis and Katrin Erk. 2020. When is a bishop not like a rook? when it's like a rabbi! multi-prototype BERT embeddings for estimating semantic relationships. In *Proceedings of the 24th Conference on Computational Natural Language Learning*, pages 227–244, Online. Association for Computational Linguistics.

Zihang Dai, Zhilin Yang, Yiming Yang, Jaime Carbonell, Quoc Le, and Ruslan Salakhutdinov. 2019. Transformer-XL: Attentive language models beyond a fixed-length context. In *Proceedings of the 57th Annual Meeting of the Association for Computational Linguistics*, pages 2978–2988, Florence, Italy. Association for Computational Linguistics.

Jacob Devlin, Ming-Wei Chang, Kenton Lee, and Kristina Toutanova. 2019. BERT: Pre-training of deep bidirectional transformers for language understanding. In *Proceedings of the 2019 Conference of the North American Chapter of the Association for Computational Linguistics: Human Language Technologies, Volume 1 (Long and Short Papers)*, pages 4171–4186, Minneapolis, Minnesota. Association for Computational Linguistics.

Matthias Fey and Jan E. Lenssen. 2019. Fast graph representation learning with PyTorch Geometric. In *ICLR Workshop on Representation Learning on Graphs and Manifolds*.

Will Hamilton, Zhitao Ying, and Jure Leskovec. 2017. Inductive representation learning on large graphs. In *Advances in Neural Information Processing Systems*, volume 30, pages 1024–1034. Curran Associates, Inc.

Jeremy Howard and Sebastian Ruder. 2018. Universal language model fine-tuning for text classification. In *Proceedings of the 56th Annual Meeting of the Association for Computational Linguistics (Volume 1: Long Papers)*, pages 328–339, Melbourne, Australia. Association for Computational Linguistics.

Eric Huang, Richard Socher, Christopher Manning, and Andrew Ng. 2012. Improving word representations via global context and multiple word prototypes. In *Proceedings of the 50th Annual Meeting of the Association for Computational Linguistics (Volume 1: Long Papers)*, pages 873–882, Jeju Island, Korea. Association for Computational Linguistics.

Luyao Huang, Chi Sun, Xipeng Qiu, and Xuanjing Huang. 2019. GlossBERT: BERT for word sense disambiguation with gloss knowledge. In *Proceedings of the 2019 Conference on Empirical Methods in Natural Language Processing and the 9th International Joint Conference on Natural Language Processing (EMNLP-IJCNLP)*, pages 3509–3514, Hong Kong, China. Association for Computational Linguistics.

Ignacio Iacobacci and Roberto Navigli. 2019. LSTMEmbed: Learning word and sense representations from a large semantically annotated corpus with long short-term memories. In *Proceedings of the 57th Annual Meeting of the Association for Computational Linguistics*, pages 1685–1695, Florence, Italy. Association for Computational Linguistics.

Thomas N. Kipf and Max Welling. 2017. Semi-Supervised Classification with Graph Convolutional Networks. In *Proceedings of the 5th International Conference on Learning Representations*, ICLR '17.

Sawan Kumar, Sharmistha Jat, Karan Saxena, and Partha Talukdar. 2019. Zero-shot word sense disambiguation using sense definition embeddings. In *Proceedings of the 57th Annual Meeting of the Association for Computational Linguistics*, pages 5670–5681, Florence, Italy. Association for Computational Linguistics.

Yoav Levine, Barak Lenz, Or Dagan, Ori Ram, Dan Padnos, Or Sharir, Shai Shalev-Shwartz, Amnon Shashua, and Yoav Shoham. 2020. SenseBERT: Driving some sense into BERT. In *Proceedings of the 58th Annual Meeting of the Association for Computational Linguistics*, pages 4656–4667, Online. Association for Computational Linguistics.

Jiwei Li and Dan Jurafsky. 2015. Do multi-sense embeddings improve natural language understanding? In *Proceedings of the 2015 Conference on Empirical Methods in Natural Language Processing*, pages 1722–1732, Lisbon, Portugal. Association for Computational Linguistics.

Robert Logan, Nelson F. Liu, Matthew E. Peters, Matt Gardner, and Sameer Singh. 2019. Barack's wife hillary: Using knowledge graphs for fact-aware language modeling. In *Proceedings of the 57th Annual Meeting of the Association for Computational Linguistics*, pages 5962–5971, Florence, Italy. Association for Computational Linguistics.

Oren Melamud, Jacob Goldberger, and Ido Dagan. 2016. context2vec: Learning generic context embedding with bidirectional LSTM. In *Proceedings of The 20th SIGNLL Conference on Computational Natural Language Learning*, pages 51–61, Berlin, Germany. Association for Computational Linguistics.

Stephen Merity, Caiming Xiong, James Bradbury, and Richard Socher. 2016. Pointer Sentinel Mixture Models. *CoRR*, abs/1609.07843.

Tomas Mikolov, Martin Karafiát, Lukás Burget, Jan Cernocký, and Sanjeev Khudanpur. 2010. Recurrent neural network based language model. In *INTERSPEECH*, pages 1045–1048. ISCA.

George A. Miller. 1995. Wordnet: A lexical database for english. *Communications of the ACM*, 38:39–41.

George A. Miller, Claudia Leacock, Ee Tengi, and Ross T. Bunker. 1993. A semantic concordance. In *Proceedings ARPA Human Language Technology Workshop*, pages 303–308.

Andrea Moro, Alessandro Raganato, and Roberto Navigli. 2014. Entity linking meets word sense disambiguation: a unified approach. *Transactions of the Association for Computational Linguistics*, 2:231–244.

Arvind Neelakantan, Jeevan Shankar, Alexandre Passos, and Andrew McCallum. 2014. Efficient non-parametric estimation of multiple embeddings per word in vector space. In *Proceedings of the 2014 Conference on Empirical Methods in Natural Language Processing (EMNLP)*, pages 1059–1069, Doha, Qatar. Association for Computational Linguistics.

Matthew E. Peters, Mark Neumann, Mohit Iyyer, Matt Gardner, Christopher Clark, Kenton Lee, and Luke Zettlemoyer. 2018. Deep contextualized word representations. In *Proceedings of the 2018 Conference of the North American Chapter of the Association for Computational Linguistics: Human Language Technologies, Volume 1 (Long Papers)*, pages 2227–2237, New Orleans, Louisiana. Association for Computational Linguistics.

Alessandro Raganato, Claudio Delli Bovi, and Roberto Navigli. 2017. Neural sequence learning models for word sense disambiguation. In *Proceedings of the 2017 Conference on Empirical Methods in Natural Language Processing*, pages 1156–1167, Copenhagen, Denmark. Association for Computational Linguistics.

Sascha Rothe and Hinrich Schütze. 2015. AutoExtend: Extending word embeddings to embeddings for synsets and lexemes. In *Proceedings of the 53rd Annual Meeting of the Association for Computational Linguistics and the 7th International Joint Conference on Natural Language Processing (Volume 1: Long Papers)*, pages 1793–1803, Beijing, China. Association for Computational Linguistics.

Bianca Scarlini, Tommaso Pasini, and Roberto Navigli. 2020a. Sensembert: Context-enhanced sense embeddings for multilingual word sense disambiguation. *Proceedings of the AAAI Conference on Artificial Intelligence*, 34(05):8758–8765.

Bianca Scarlini, Tommaso Pasini, and Roberto Navigli. 2020b. With more contexts comes better performance: Contextualized sense embeddings for all-round word sense disambiguation. In *Proceedings of the 2020 Conference on Empirical Methods in Natural Language Processing (EMNLP)*, pages 3528–3539, Online. Association for Computational Linguistics.

Ashish Vaswani, Noam Shazeer, Niki Parmar, Jakob Uszkoreit, Llion Jones, Aidan N Gomez, Łukasz Kaiser, and Illia Polosukhin. 2017. Attention is All you Need. *Advances in Neural Information Processing Systems*, 30:5998–6008.

Petar Veličković, Guillem Cucurull, Arantxa Casanova, Adriana Romero, Pietro Liò, and Yoshua Bengio. 2018. Graph attention networks. In *ICLR*.

Loïc Vial, Benjamin Lecouteux, and Didier Schwab. 2018. UFSAC: Unification of Sense Annotated Corpora and Tools. In *Language Resources and Evaluation Conference (LREC)*, Miyazaki, Japan.

.

Logical Story Representations via FrameNet + Semantic Parsing

Lane Lawley and **Lenhart Schubert**
University of Rochester
Department of Computer Science
{llawley,schubert}@cs.rochester.edu

Abstract

We propose a means of augmenting FrameNet parsers with a formal logic parser to obtain rich semantic representations of events. These schematic representations of the frame events, which we call Episodic Logic (EL) schemas, abstract constants to variables, preserving their types and relationships to other individuals in the same text. Due to the temporal semantics of the chosen logical formalism, all identified schemas in a text are also assigned temporally bound "episodes" and related to one another in time. The semantic role information from the FrameNet frames is also incorporated into the schema's type constraints. We describe an implementation of this method using a neural FrameNet parser, and discuss the approach's possible applications to question answering and open-domain event schema learning.

1 Introduction

Story understanding requires deep, non-textual representations of textual information. The human brain, neural language models, and formal logic engines all transduce textual input into some other format in order to perform semantic tasks on that input. While formal logical representations of language admit more reliable and explainable inference procedures on text than, for example, the vector representations used by transformers, they suffer from characteristic brittleness when attempting to parse the true logical *meaning* of text: paraphrases and idioms stymie the logical capture of true semantics at best, and actively lead to incorrect understanding at worst.

The FrameNet project (Baker et al., 1998) attempts to provide a taxonomy of event "frames" (sometimes also called "schemas" or "scripts"), including their actors and objects, that one might observe in the real world, and thus in texts discussing the real world. These frames are not tied to any one means of expression: many different constructions,

```
EPI-SCHEMA ((?X_C (COMPOSITE-SCHEMA.PR ?X_D)) ** ?E)
    :ROLES
        !R1  (?X_A FRIEND.N)
        !R2  (?X_A (PERTAIN-TO ?X_B))
        !R3  (?X_B AGENT.N)
        !R4  (?X_C MOM.N)
        !R5  (?X_C (PERTAIN-TO ?X_B))
        !R6  (?X_C MOTION-THEME.N)
        !R7  (?X_C INGESTION-INGESTOR.N)
        !R8  (?X_D HOUSE.N)
        !R9  (?X_D (PERTAIN-TO ?X_A))
        !R10 (?X_D MOTION-GOAL.N)
        !R11 (?X_E FOOD.N)
        !R12 (?X_E INGESTION-INGESTIBLES.N)

    :STEPS
        ?E1 (?X_C MOTION-GO.1.V ?X_D)
        ?E2 (?X_C INGESTION-EAT.2.V ?X_E)

    :EPISODE-RELATIONS
        !W2 (?E1 BEFORE ?E2)
```

Figure 1: An example of an Episodic Logic schema representing the story *"Jenny's mom went to her friend's house. She ate food there."* Constants in this story, such as "Jenny", have been abstracted to variable names, creating a general schema form of the story, but the original story constants may be re-bound to these variables at any time. Noun predicates taken from single story tokens, e.g. FRIEND.N, are color-coded with their variables. Noun and verb predicates obtained from FrameNet matches are underlined, and prefixed with the name of the FrameNet frame before the hyphen. Additional information on the syntax and semantics of the schema is given by Lawley et al. (2021).

Proceedings of the Workshop on Dimensions of Meaning: Distributional and Curated Semantics (DistCurate 2022), pages 19 - 23
July 14, 2022

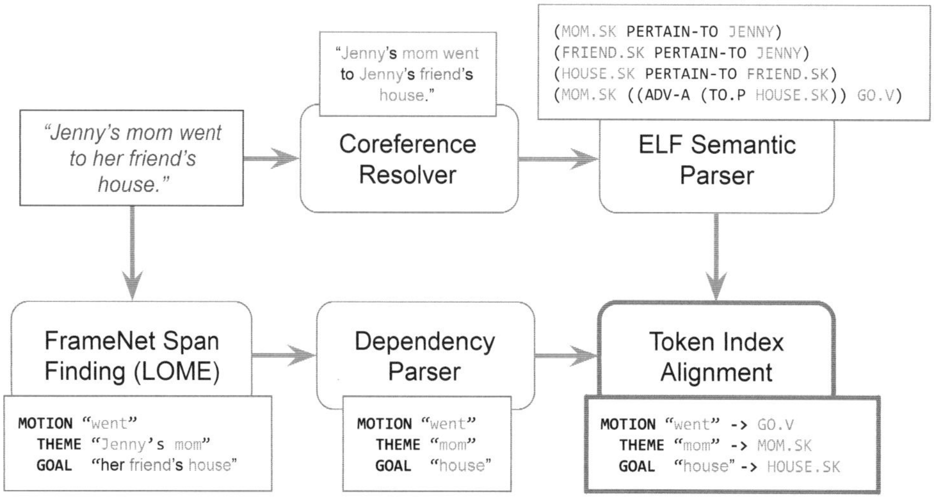

Figure 2: The architecture of the system. Raw story text is fed along two tracks: the logical-semantic parsing track, shown along the top, and the FrameNet parsing track, shown along the bottom. The FrameNet text spans are reduced to direct object tokens and correlated with logical individuals in the ELF parse via token index matching.

e.g. "she wolfed down the meal" and "she ate her food", can express the same frame, e.g. "ingestion". These frames are constructed manually, however, rather than learned automatically from texts, and are defined in terms of natural language rather than a more manipulable representation. FrameNet parsing of text generally consists of the mapping of spans of text to FrameNet roles; these text spans, being natural language, are difficult to manipulate programmatically and draw inferences from.

In this paper, we present a means of producing expressive, semantically manipulable, formal logical "schema" representations of stories using a state-of-the-art FrameNet parsing system, LOME (Xia et al., 2021), as a jumping-off point. By augmenting FrameNet parses with logical semantic representations of the text, we obtain schema-like story representations that mitigate both the brittleness inherent to literal semantic parsing and the difficulty of manipulation inherent to natural language frames. We also discuss the potential application of these representations to the task of automatically acquiring event schema knowledge from natural text corpora.

2 Semantic Representation

The semantic representation we provide is based on Episodic Logic (EL) (Hwang and Schubert, 1993), a formal logical representation of language that enables efficient inference while maintaining a surface resemblance to the English language. One feature of EL that is well suited to story representation is its *characterizing* operator, `**`, which relates an Episodic Logic Formula (ELF) to an *episode*. Informally, `(`ϕ` ** E)` means that `E` is "an episode of" some formula ϕ, e.g., in `((?X_C MOTION-GO.1.V ?X_D) ** ?E1)`, `?E1` is an episode of `?X_C` going to `?X_D` (cf. the first step in Figure 1; in schemas the `**` operator is left implicit). These episodes, characterized by formulas derived from sentences, have temporal bounds, and can be related to each other in time using relations derived from the Allen Interval Algebra (Allen, 1983). Episodes are first-class individuals in Episodic Logic, and may be used as arguments to predicates, such as in the temporal relation formula `(E1 BEFORE E2)`.

ELFs, like those seen in the schema in Figure 1, often have predicates derived from nouns or verbs. For example, the first role condition in Figure 1, the schema ELF `(?X_A FRIEND.N)`, asserts that the variable `?X_A` satisfies the predicate `FRIEND.N` (and, as stated in the next role condition, `?X_A` "pertains to" `?X_B`, i.e. `?X_A` is a friend *of* Jenny). The first step of the same schema, the ELF `(?X_C MOTION-GO.1.V ?X_D)`, can be read as a subject-verb-object verb phrase, where the arguments to the verb predicate, `MOTION-GO.1.V`, are the variables `?X_C` and `?X_D`.

2.1 EL Schemas

To represent frames identified by the FrameNet parser, as well as the story as a whole, we use the schema system built atop EL by Lawley et al. (2021). An example schema, produced by the system presented in this paper, is shown in Figure 1. This schema format allows declaration of entity types, and of relationships between entities, via EL propositions in the `Roles` section. The `Steps` section contains ELFs, and their characterizing episodes, for the schema's constituent events. These episodes are related in the `Episode-relations` section, and the entire schema may itself be embedded by the ELF formula known as its *header*, visible at the top of the schema, and characterizing an episode itself.

The EL schema framework we use allows for other section types, such as goals, preconditions, and postconditions, and was designed as part of a larger schema acquisition project. In this work, however, we primarily make use of the `Roles`, `Steps`, and `Episode-relations` sections for frame and story representation.

3 Architecture

Our system's architecture, illustrated in Figure 2, is divided into two main information pipelines: the EL track, responsible for semantic parsing, and the FrameNet track, responsible for frame identification and span selection. The information from both of these pipelines is unified into a final schematic representation at the end using token indices from the input text.

3.1 EL Track

To produce an EL semantic parse of the story, we first perform span mapping on the input text using the AllenNLP coreference resolver (Gardner et al., 2017). Co-referring token indices are saved, and story sentences are then converted into ELFs by first parsing them into ULF—an underspecified variant of EL (Kim and Schubert, 2019)—and then processing the ULFs into full ELFs by converting grammatical tense information into temporal relations and scoping quantifiers. More information on the ELF parser can be found in (Lawley et al., 2021).

Coreference resolution on the ELFs is performed by cross-referencing the token index clusters with token index tags placed on individuals in the EL parse. Co-referring individuals in the EL parse are then combined into one individual and substitutions are made throughout the parse.

3.2 FrameNet Track

To identify basic behavioral frames invoked by the raw text, we make use of the LOME information extraction system (Xia et al., 2021). LOME outputs invoked frames, and text spans that fill frame roles, as CONCRETE data files. Once we extract the invoked frames and text spans, we perform a syntactic dependency parse on the input text using spaCy (Honnibal and Montani, 2017) and identify the first token in each span with a `NSUBJ`, `DOBJ`, or `POBJ` tag. This allows any span of text containing tokens for multiple individuals, e.g. *her friend's house*, to be reduced to, e.g., *house*, which will be the token used to identify the logical individual in the EL parse during the alignment phase.

3.3 Token Index Alignment and Schema Formation

To represent the identified FrameNet frames as EL formulas, the text spans that fill the semantic roles for each frame must first be bound to logical individuals. After the dependency parser identifies the token to cross-reference with the EL parse, the noun predicate with the same token index is retrieved from the EL parse, and the individual satisfying that predicate is identified as the bound value for the frame role.

The verb that invoked the frame is identified in a similar fashion, and a schema is created with that verb's formula from the EL parse as its header, and with the names of the FrameNet semantic roles applied to the relevant individuals as semantic types in the new schema's `Roles` section. When multiple frames are converted to schemas in this way, they may all be embedded together in a *composite schema*, such as the one shown in Figure 1, with their header formulas as steps and with each of their inner type constraints shown in the composite schema's `Roles` section for clarity. This composite schema forms our final semantic representation of the story.

4 Discussion

The goal of our representation, and of semantic story representations in general, is to enable a variety of reasoning tasks. As the quality of the frames identified by LOME has already been evaluated by Xia et al. (2021), we do not re-evaluate quality after transducing those frames into EL schemas. Below, we discuss two interesting potential applications of this representation: question answering and event schema acquisition.

4.1 Applications

4.1.1 Question Answering

Episodic Logic has been used for question answering (Morbini and Schubert, 2009), as has its underspecified variant, ULF (Platonov et al., 2020). EL formulas can be *unified* with one another, binding variables in one formula to constants or variables in another. Many questions about events or types can be formulated as EL propositions with variables to be bound to potential answers. For example, to answer the question of whose house the mom went to in the story represented in Figure 1, we could create the question formulas with new variables for the house and its owner: `(?X_C MOTION-GO.1.V ?house)` and `(?house (PERTAIN-TO ?who))`. The only valid unification of these formulas with the story binds the house `?X_D` to `?house` and the friend `?X_A` to `?who`. FrameNet-based representations make answerable questions somewhat paraphrase-resistant, as well: "whose house did the mom run off to?" would invoke the same frame.

This form of question answering may also be used for semantic information retrieval based on multiple separate type, relational, and event occurrence constraints, for example, finding sets of stories where a person buys something edible at a store.

4.1.2 Schema Learning

When information about stereotypical situations is packaged up into event schemas, those schemas may be partially matched to new stories, and inferences may then be drawn from the unmatched pieces of those schemas: upon observing someone sitting down at a restaurant, for example, you might infer that they would then receive a menu.

The event schema syntax we use, taken from (Lawley et al., 2021), was conceived as part of a system for learning rich, logical event schemas from texts by using a set of simple behavioral *protoschemas*—concepts children are familiar with, like asking for assistance with a task or eating food to alleviate hunger—to bootstrap the acquisition of more complex schemas. We believe that our conversion of identified FrameNet frames to canonicalized logical formulas could aid this process: many FrameNet frames resemble simple behavioral protoschemas, and a mapping between them has been already been employed for existing schema learning work based on protoschemas (Lawley and Schubert, 2022).

4.2 Limitations

While our system produces useful representations, extant Episodic Logic parsing software, especially ULF parsing, is still somewhat error-prone. Work on EL parsing is ongoing, and notably includes an application of the cache transition parsing system developed by Peng et al. (2018) to ULF parsing (Kim, 2019), which is the initial step in converting English text into a logical form.

We also note that we do not leverage the full schema syntax of Lawley et al. (2021), and in particular have not added stated goals, preconditions, and postconditions from FrameNet frames into the relevant sections from that schema system. This is due, in large part, to the lack of availability of those particular semantic roles in current FrameNet parses.

Finally, we note that our system was developed using only stories from the ROCstory corpus (Mostafazadeh et al., 2016), and that grammatically and conceptually complex texts may require additional parsing techniques; better parser performance; a larger corpus of schemas, with the initial hand-created basic schemas expanded through schema learning; or any subset of these.

5 Conclusion

We have presented a system for obtaining rich, formal logic-based, schema-like representations of stories from text by combining the frame identification power of LOME and FrameNet with the semantic representation power of Episodic Logic schemas. We showed that these representations normalize language into propositions based on semantic frames; model type, relational, and temporal constraints; and allow for hierarchical nesting of situations. Finally, we discussed their potential application, in future work, to tasks that neither FrameNet nor EL parsing alone is trivially capable of, such as paraphrase-resistant question answering, information retrieval, and automatic acquisition of event schemas from text, to which this system has already been applied.

References

James F. Allen. 1983. Maintaining knowledge about temporal intervals. *Commun. ACM*, 26(11):832–843.

Collin F. Baker, Charles J. Fillmore, and John B. Lowe. 1998. The berkeley framenet project. In *Proceedings of the 36th Annual Meeting of the Association for Computational Linguistics and 17th International Conference on Computational Linguistics - Volume 1*, ACL '98/COLING '98, pages 86–90. Association for Computational Linguistics.

Matt Gardner, Joel Grus, Mark Neumann, Oyvind Tafjord, Pradeep Dasigi, Nelson F. Liu, Matthew Peters, Michael Schmitz, and Luke S. Zettlemoyer. 2017. Allennlp: A deep semantic natural language processing platform.

Matthew Honnibal and Ines Montani. 2017. spaCy 2: Natural language understanding with Bloom embeddings, convolutional neural networks and incremental parsing. To appear.

Chung Hee Hwang and Lenhart K Schubert. 1993. Episodic logic: A situational logic for natural language processing. *Situation Theory and its Applications*, 3:303–338.

Gene Kim and Lenhart Schubert. 2019. A type-coherent, expressive representation as an initial step to language understanding. In *Proceedings of the 13th International Conference on Computational Semantics*, Gothenburg, Sweden. Association for Computational Linguistics.

Gene Louis Kim. 2019. Towards parsing unscoped episodic logical forms with a cache transition parser. In *the Poster Abstracts of the Proceedings of the 32nd International Conference of the Florida Artificial Intelligence Research Society*.

Lane Lawley, Benjamin Kuehnert, and Lenhart Schubert. 2021. Learning general event schemas with episodic logic. In *Proceedings of the 1st and 2nd Workshops on Natural Logic Meets Machine Learning (NALOMA)*, pages 1–6, Groningen, the Netherlands (online). Association for Computational Linguistics.

Lane Lawley and Lenhart Schubert. 2022. Mining logical event schemas from pre-trained language models. In *Proceedings of the 60th Annual Meeting of the Association for Computational Linguistics: Student Research Workshop*, Dublin, Ireland. Association for Computational Linguistics. To appear.

Fabrizio Morbini and Lenhart Schubert. 2009. Evaluation of Epilog: A reasoner for Episodic Logic. In *Proceedings of the Ninth International Symposium on Logical Formalizations of Commonsense Reasoning*, Toronto, Canada.

Nasrin Mostafazadeh, Nathanael Chambers, Xiaodong He, Devi Parikh, Dhruv Batra, Lucy Vanderwende, Pushmeet Kohli, and James Allen. 2016. A corpus and cloze evaluation for deeper understanding of commonsense stories. In *Proceedings of the 2016 Conference of the North American Chapter of the Association for Computational Linguistics: Human Language Technologies*, pages 839–849, San Diego, California. Association for Computational Linguistics.

Xiaochang Peng, Daniel Gildea, and Giorgio Satta. 2018. AMR parsing with cache transition systems. In *Proceedings of the National Conference on Artificial Intelligence (AAAI-18)*.

Georgiy Platonov, Lenhart Schubert, Benjamin Kane, and Aaron Gindi. 2020. A spoken dialogue system for spatial question answering in a physical blocks world. In *Proceedings of the 21th Annual Meeting of the Special Interest Group on Discourse and Dialogue*, pages 128–131, 1st virtual meeting. Association for Computational Linguistics.

Patrick Xia, Guanghui Qin, Siddharth Vashishtha, Yunmo Chen, Tongfei Chen, Chandler May, Craig Harman, Kyle Rawlins, Aaron Steven White, and Benjamin Van Durme. 2021. LOME: Large ontology multilingual extraction. In *Proceedings of the 16th Conference of the European Chapter of the Association for Computational Linguistics: System Demonstrations*, pages 149–159, Online. Association for Computational Linguistics.

Comparing Distributional and Curated Approaches for Cross-lingual Frame Alignment

Collin F. Baker Michael Ellsworth Miriam R. L. Petruck **Arthur Lorenzi**
International Computer Science Institute Universidade Federal de Juiz de Fora
Berkeley, California Minas Gerais, Brazil
{collinb,infinity,miriamp} arthur.lorenzi@gmail.com
@icsi.berkeley.edu

Abstract

Despite advances in statistical approaches to the modeling of meaning, many questions about the ideal way of exploiting both knowledge-based (e.g., FrameNet, WordNet) and data-based methods (e.g., BERT) remain unresolved. This workshop focuses on these questions with three session papers that run the gamut from highly distributional methods (Lekkas et al., 2022), to highly curated methods (Gamonal, 2022), and techniques with statistical methods producing structured semantics (Lawley and Schubert, 2022).

In addition, we begin the workshop with a small comparison of cross-lingual techniques for frame semantic alignment for one language pair (Spanish and English). None of the distributional techniques consistently aligns the 1-best frame match from English to Spanish, all failing in at least one case. Predicting which techniques will align which frames cross-linguistically is not possible from any known characteristic of the alignment technique or the frames. Although distributional techniques are a rich source of semantic information for many tasks, at present curated, knowledge-based semantics remains the only technique that can consistently align frames across languages.

1 Introduction to the Workshop

Broadly speaking, research in computational linguistics encompasses two main streams: (1) work that relies primarily on operationalizing prior knowledge about language and its use, such as rule-based parsers (Bender et al., 2002), scripts (Schank and Abelson, 1977), planning, scenarios, scripts for virtual assistants, and FrameNet (**FN**) frames (Ruppenhofer et al., 2016), as well as lexical databases like WordNet (Fellbaum, 1998), VerbNet (Kipper et al., 2000), and PropBank (Palmer et al., 2005), among others; and (2) work that seeks to derive knowledge directly from data (text, speech, and increasingly vision) with unsupervised (or distantly supervised) methods, which are distributional and frequency-based, in linguistics (Biber et al., 2020), cognitive science (Xu and Xu, 2021), and computational linguistics, notably vector embeddings like BERT (Devlin et al., 2019). They are often complementary; e.g. Kuznetsov and Gurevych (2018) combine POS tagging and lemmatization to improve vector embeddings and Qian et al. (2021) combine syntactic knowledge with neural language models to improve accuracy.

Despite great advances in statistical approaches, many questions remain unresolved:

- What are the strengths and limitations of each approach?
- Is extracting different types of knowledge from text/speech possible by one and not the other? Why?
- How well can each represent relations and support reasoning over text?
- What factors limit progress of each approach?
- Would combining the two approaches solve all the problems?

These issues are as pertinent today as they were nearly 30 years ago at "The Balancing Act: Combining Symbolic and Statistical Approaches to Language" (McDonald, 1994). The goal of this workshop is to encourage reporting of research bearing on these issues; we will hear three such papers (listed below), in addition to our own results on cross-lingual frame alignment, described in the remainder of this paper.

In "Multi-sense Language Modelling", Andrea Lekkas, Peter Schneider-Kamp and Isabelle Augenstein use pretrained embeddings and also calculate new ones, combining them with many facets of the curated WordNet lexicon. They report on extensive testing of five different system architectures against a most-frequent-sense baseline on both next word prediction and WordNet sense prediction, on both the SemCor and SemEval datasets.

Maucha Gamonal's"A Descriptive Study of

Proceedings of the Workshop on Dimensions of Meaning: Distributional and Curated Semantics (DistCurate 2022), pages 24 - 30
July 14, 2022 ©2022 Association for Computational Linguistics

Metaphors and Frames in the Multilingual Shared Annotation Task" shows how FrameNet frames can explain instances of metaphor in 50 sentences from the transcription of a TED talk that members of the respective FrameNet projects annotated in Portuguese, English, and German. The"frame shift" discussed in the paper also have implications for theories of translation. Despite progress on automatic recognition of metaphors (e.g. Veale 2016, Shutova et al. 2015, Chakrabarty et al. 2021), the kind of detail shown here is generally not retrievable computationally.

Lane Lawley and Lenhart Schubert generate"Logical Story Representations via FrameNet + Semantic Parsing". Lawley and Schubert have previously worked on learning logical representations of events ("event Logic") from simple stories (Lawley et al., 2021). In this paper, they show how semantic parsing based on FrameNet and implemented in the LOME parser (Xia et al., 2021) can add valuable information to the logical representation, allowing more precise reasoning.

As described in the rest of this paper, the introduction to the workshop presents but one example of the complex interplay between curated and distributional semantics from research at the ICSI FrameNet project: cross-linguistic frame alignment. We compared curated semantics techniques with those of unsupervised distributional ones, intentionally focusing on a very small set of data for one language pair (English and Spanish) to characterize the specific details of such comparisons.

The remainder of the paper proceeds as follows: Section 2 provides a brief overview of FrameNet; Section 3 describes the work of developing cross-lingual frame alignments; Section 4 presents the results of different methods for aligning some Spanish and English frames; and Section 5 offers concluding remarks and future directions to pursue cross-linguistic frame alignment.

2 Overview of FrameNet

FrameNet (Ruppenhofer et al., 2016) is a research and resource development project in corpus-based computational lexicography grounded in the theory of **Frame Semantics** (Fillmore, 1985).

The **semantic frame**, a script-like knowledge structure that facilitates inferencing within and across events, situations, states-of-affairs, relations, etc., is at the core of the theory (Petruck, 1996). FN defines a semantic frame in terms of its **frame elements** (FEs), or participants (and other concepts) in the scene that the frame captures; a **lexical unit** (LU) is a pairing of a lemma and a frame, characterizing that LU in terms of the frame that it evokes. The definition of a frame, represented both in prose and in structured relations between frames, is a bundle of inferences relating the frame elements whenever the frame is evoked.

Example 1 illustrates the Frame Semantics analysis for the verb **buy**, which FN defines in the `Commerce_buy` frame, with the FEs BUYER, SELLER, GOODS, and MONEY.[1]

1. Chuck BUYER **bought** a car GOODS from Jerry SELLER for $2,000 MONEY

3 Cross-Linguistic Frame Alignment

As interest in Frame Semantics (Fillmore, 1982) and the original FrameNet for English (Fillmore, 2014) grew, research groups around the world started developing FN-like resources for their languages. Such resources in many languages have made it possible to address the question of whether semantic frames are universal or merely language-specific lexical phenomena. With these databases at hand, we may operationalize the question as: To what extent can these lexical databases be aligned to form a multilingual FrameNet lexical database connecting all of the languages, while also accounting for language-specific differences and domain-specific extensions to FrameNet?

The goal of the Multilingual FrameNet (**MLFN**) project (Gilardi and Baker, 2018) was to answer this question by building a cross-linguistic database. Though this database succeeded in partially aligning frames, the question remained of how to assess the validity and utility of the alignments. Baker and Lorenzi (2020) described a database of vectors that represent alignments between pairs of frames in different languages (e.g., English-Spanish, English-Japanese, etc.).[2] Baker and Lorenzi (2020) also described developing ViToXF, a freely available visualization tool for all of the alignments.[3] The tool allows interactive exploration of the alignments between English and one of seven other languages

[1]This paper uses these typographical conventions: Frame names are in `typewriter font`; FE Names are in SMALL CAPS; and lexical units are in **boldface**.

[2]The latest release of this database is available on Github. `https://github.com/icsi-berkeley/framenet-multilingual-alignment/releases/tag/1.0.3-2`.

[3]`https://github.com/icsi-berkeley/framenet-multilingual-alignment`.

in the database. The alignments were created using 11 different methods, four (4) resource-based and seven (7) vector-based. The rest of this section characterizes these alignment methods; this is a revised version of the descriptions of alignment methods in Baker and Lorenzi (2020).

3.1 Alignment by Frame Name/ID Number

At first glance, the alignment problem might seem trivial: if other FNs have used Berkeley FrameNet (**BFN**) frame names. Can we assume that a frame in a language other than English with the same name as a BFN frame represents the same concept, and just ignore any that don't have matching names? Furthermore, some of the other resources used target-language frame names, rather than English ones, a situation that would mean aligning just the names themselves by translation between the two languages. Sometimes, the non-English language frame data also included a field for the BFN frame name or BFN frame ID, which could be used independently for alignment. Also, even when the names (or IDs) match, the non-English frame may be defined differently or have a different number of core FEs than the BFN frame. By definition, a different number of core FEs between the (English and non-English) frames amounts to different frames, so such alignments are, at best, imperfect.

3.2 Alignment by LU Translation

A second way of approaching alignment is to take all the LUs from a source language frame and find translation equivalents in the target language. If frames are equivalent across languages, we expect the translations of LUs in one source language frame to fall in the same target language frame, but the success of this method depends on the accuracy of the translations. By definition, a LU represents one sense of a lemma, a fact that should, in principle, greatly narrow the range of possible translations. However, exploiting frame information in the translation process remains a challenge.

We use The Open Multilingual WordNet (**OMWN**) (Bond and Foster, 2013) to find translation equivalents between languages. The first step is to create a mapping $S(\ell)$ from each LU in each language to a set of synsets one of which may represent its sense. That mapping requires finding OMWN synsets that contain the lemma+POS of the given LU . Let L_e and L_f be the lists of LUs in any two frames in the source language (e) and the target language (f). Equation 1 defines the matching of LUs between L_e and L_f.

$$\mathrm{m}_1(L_e, L_f) = \{a \in L_e \mid b \in L_f, \; S(a) \cap S(b) \neq \emptyset\} \quad (1)$$

To evaluate the alignment between the two frames, this function calculated three different scores (selectable in ViToXF under the name "LU translations using WordNet"). The first is a metric that considered LUs from both frames (Equation 2), but this method gives too much weight to frames containing more LUs. Avoiding this problem required breaking the alignment into two scores, accounting for the direction of alignment. Specifically, the score of the alignment from English to the target language might be different from the reverse. Equation 3 presents the formula for one of those scores. (Simply switching the two arguments will obtain the other score.)

$$\mathrm{s}_1(L_e, L_f) = \frac{|f(L_e, L_f)| + |f(L_f, L_e)|}{|L_e| + |L_f|} \quad (2)$$

$$\mathrm{s}_2(L_e, L_f) = \frac{|f(L_e, L_f)|}{|L_e|} \quad (3)$$

We also explored an alternative scoring method based on synsets rather than LUs (by selecting "Synset count" in ViToXF). Equation 4 defines the matching set in this case, with the scores calculated in a manner similar to that of Equation 3.

$$\mathrm{m}_2(L_e, L_f) = \bigcup_{a \in L_e} S(a) \cap \bigcup_{b \in L_f} S(b) \quad (4)$$

3.3 Alignment by Frame Element Similarity

Recalling that frames are defined in terms of the entailments of their FEs, for two frames to be the same across languages, they must minimally have the same number and type of FEs. Some FrameNets, like Spanish FN and Japanese FN, simply used the same FEs that BFN named and defined; that is, the names and definitions of FEs are identical to those of the English resources. Others, e.g., Chinese, translated the names and the definitions into the target language or created completely new ones in the target language. These cases required aligning the FEs according to the proximity of the names and definitions from the two languages in a shared vector space. French created FE names and definitions in English, although many of those frames do not correspond to those in BFN. Swedish FN used FE names in English and adopted the BFN definitions. Both the Brazilian Portuguese and German FN projects include FEs in a mixture of English and the target language. In these last two cases, developing the alignment required grouping the FEs according to the language of their name (English or the target language), calculating the similarity separately for the FEs in each language, and then combining the scores.[4]

3.4 Alignment by Distributional Similarity of Lexical Units

Another approach uses cross-lingual word embeddings to find alignments; this appears in ViToXF under the options “LU translations using MUSE” and “MUSE centroid similarity”. Currently, ViToFX is based on fastText word embeddings from various languages trained on Wikipedia data and aligned to a single vector space (Bojanowski et al., 2017). The spaces were aligned by an unsupervised adversarial approach, where the discriminator tries to predict the embedding origin and the generator aims to create transformations that the former cannot accurately classify (Lample et al., 2018). The transformed fastText vectors of many languages mapped to English space are available in the MUSE library.[5] MLFN uses these pre-trained cross-lingual word embeddings for two different scoring techniques. The first, "LU translations using MUSE" (like those in Section 3.2), uses the word embeddings as a way to obtain translation equivalents. We define $n(\vec{v}, k, t)$, the k-neighborhood of $\vec{v}$ in the target language with cosine similarity greater than t. Equation 5 defines the alignment score between a pair of frames given their LU lists L_e and L_f.

$$s_3(L_e, L_f) = \frac{|\{a \in L_e \mid b \in L_f, \vec{v}(b) \in n(\vec{v}(a), k, t)\}|}{|L_e|} \quad (5)$$

The second distributional technique, “MUSE centroid similarity”, calculates the alignment between two frames by finding the average vector of their LUs vectors (i.e. the centroid vector of each frame) and computing the cosine similarity of those centroids, like Sikos and Padó (2018).

4 Results

To evaluate the alignments created by the various techniques described in Section 3 (above), FN researchers defined a set of "gold-standard" frame alignments for a small set of frames from Spanish FrameNet (SFN) aligned to English frames from BFN.[6] We determined gold-standard frame matches manually by comparing all of the information associated with the frames of each language, including frame definition, frame elements, lexical units with their translations, and frame relations (if any). Since the time for a manual review precludes comparing all frames to all other frames, we only considered those frames with lexical translation overlap.[7]

For each gold-standard alignment, we examined the full set of alignment techniques provided by ViToXF for the SFN frame. With ViToXF, each technique will align a SFN frame to different list of BFN frames, and each such pairing will have a score. The techniques have very different scores, even when normalized, so the best way to compare techniques is by how they order the proposed BFN frames to align. In what follows, we simplify the evaluation of a technique to the relative rank (*1st*, *2nd*, etc.) of the gold-standard BFN frame.

Table 1 compares alignments of five SFN frames (ending in *.es*) with those of BFN (ending *.en*). The first two rows show the gold standard alignments; in four cases, the frames have the same name in both languages. However, SFN `Motion_manner` corresponds most closely with BFN’s `Self_motion`. The other four rows show the rank (1st, 2nd, etc.) of the English gold standard frame in the output of each of four alignment algorithms:

1. Proportion of matching core FE names or IDs
2. WordNet synset count (mapped from Spanish to English synsets)
3. MUSE LU centroid similarity
4. Average core FE name/definition similarity (using MUSE vectors)

Note that the first two of the above-mentioned algorithms are entirely based on curated resources, the

[4]The process used Michal Danilak’s python library for language recognition `https://pypi.org/project/langdetect/`

[5]https://github.com/facebookresearch/MUSE

[6]The data derive from SFN (Subirats-Rüggeberg and Petruck, 2003) and V.1.7 of BFN (Ruppenhofer et al., 2016).

[7]Some of these results were also described in the Call for Papers for this workshop.

Spanish Frames	Desiring.es	Motion manner.es	Performers and roles.es	Similarity.es	Activity finish.es
Gold standard English match	Desiring.en	Self-motion.en	Performers and roles.en	Similarity.en	Activity finish.en
Matching core FE names/IDs	1st	No match	1st	1st	1st-8th
WN synset count (es-> en)	1st	1st	1st	1st	2nd-3rd
MUSE LU centroid similarity	1st	2nd	2nd	1st	2nd
Average core FE MUSE similarity	1st	>10th	1st	1st	1st

Table 1: Rank of Gold-standard Frame Match by Alignment Method

third is purely distributional, and the last of these combines the two approaches.

All the techniques show promise in accurately aligning certain frames and all perform less well on the inexact match, i.e., SFN `Motion_manner` to BFN `Self_motion`. Unlike the English frame, the SFN frame does not allow complex path information, although many LUs in the SFN frame have translation equivalents in BFN `Self_motion` (e.g. Spanish *correr*.v -> English *run*.v). Also, no single technique ranks the gold standard as the strongest match for all of the listed frames. SFN `Desiring` and `Similarity` align correctly by all of the techniques listed; in contrast, SFN `Activity_finish` only aligns unambiguously using just one technique, i.e., Average core FE MUSE similarity.

At least based on this limited data, it is not possible to predict which techniques will do a good job of aligning which frames cross-linguistically from any known characteristic of the alignment technique or the frames. This is just one example of the complex questions involved in comparing different approaches to alignment.

5 Concluding Remarks and Future Work

This paper has introduced the workshop exploring the strengths and weaknesses of different techniques for modeling meaning. Specifically, the research of the Berkeley FrameNet group has compared distributional approaches and curated approaches for cross-lingual frame alignment, illustrating the results from four different alignment techniques for five Spanish FrameNet and BFN frames, finding that no distributional technique reliably predicts the gold-standard alignment.

This initial study on a small set of frames in only two languages is suggestive, and points to the need for wider exploration of techniques for aligning lexical units and frames across languages in frame-based resources. We look forward to further development of hybrid semantic representations combining the advantages of distributional and curated semantic techniques, both for the alignment task and a wider range of applications.

6 Acknowledgements

We are especially grateful to the FrameNet research groups that have provided their data for this study. We hope to add more soon.

- Spanish FN (Subirats, 2009)
- SALSA (Burchardt et al., 2006)
- Japanese FrameNet (Ohara et al., 2004)
- Chinese FN (You and Liu, 2005)
- FrameNet Brasil (Torrent et al., 2018)
- Swedish FN++ (Borin et al., 2010)
- French FN (Candito et al., 2014)
- Dutch FN (Vossen et al., 2018)

The U.S. National Science Foundation has supported the research on frame alignment and the creation of ViToXF under Grant No. (1629989). Any opinions, findings, and conclusions or recommendations expressed in this material are those of the authors and do not necessarily reflect the views of the National Science Foundation.

References

Collin F. Baker and Arthur Lorenzi. 2020. Exploring crosslinguistic frame alignment. In *Proceedings of the International FrameNet Workshop 2020: Towards a Global, Multilingual FrameNet*, pages 77–84, Marseille, France. European Language Resources Association.

Emily M. Bender, Dan Flickinger, and Stephan Oepen. 2002. The Grammar Matrix: An open-source starter-kit for the rapid development of cross-linguistically consistent broad-coverage precision grammars. In *Proceedings of the Workshop on Grammar Engineering and Evaluation at the 19th International Conference on Computational Linguistics*, pages 8–14, Taipei, Taiwan.

Douglas Biber, Jesse Egbert, and Daniel Keller. 2020. Reconceptualizing register in a continuous situational space. *Corpus Linguistics and Linguistic Theory*, 16(3):581–616.

Piotr Bojanowski, Edouard Grave, Armand Joulin, and Tomas Mikolov. 2017. Enriching word vectors with subword information. *Transactions of the Association for Computational Linguistics*, 5:135–146.

Francis Bond and Ryan Foster. 2013. Linking and extending an Open Multilingual Wordnet. In *Proceedings of the 51st Annual Meeting of the Association for Computational Linguistics (Volume 1: Long Papers)*, pages 1352–1362, Sofia, Bulgaria. Association for Computational Linguistics.

Lars Borin, Dana Danélls, Markus Forsberg, Dimitrios Kokkinakis, and Maria Toporowska Gronostaj. 2010. The Past Meets the Present in Swedish FrameNet++. In *Proceedings of EURALEX 14*, pages 269–281. EURALEX.

Aljoscha Burchardt, Katrin Erk, Anette Frank, Andrea Kowalski, and Sebastian Padó. 2006. The SALSA Corpus: a German Corpus Resource for Lexical Semantics. In *Proceedings of LREC 2006, Workshop on Semantically Annotated Corpora*, Genoa, Italy. ELRA.

Marie Candito, Pascal Amsili, Lucie Barque, Farah Benamara, Gaël De Chalendar, Marianne Djemaa, Pauline Haas, Richard Huyghe, Yvette Yannick Mathieu, Philippe Muller, et al. 2014. Developing a french framenet: Methodology and first results. In *LREC-The 9th edition of the Language Resources and Evaluation Conference*.

Tuhin Chakrabarty, Debanjan Ghosh, Adam Poliak, and Smaranda Muresan. 2021. Figurative language in recognizing textual entailment. In *Findings of the Association for Computational Linguistics: ACL-IJCNLP 2021*, pages 3354–3361, Online. Association for Computational Linguistics.

Jacob Devlin, Ming-Wei Chang, Kenton Lee, and Kristina Toutanova. 2019. BERT: Pre-training of deep bidirectional transformers for language understanding. In *Proceedings of the 2019 Conference of the North American Chapter of the Association for Computational Linguistics: Human Language Technologies, Volume 1 (Long and Short Papers)*, pages 4171–4186, Minneapolis, Minnesota. Association for Computational Linguistics.

Christane Fellbaum, editor. 1998. *WordNet. An electronic lexical database*. MIT Press, Cambridge/Mass.

Charles J. Fillmore. 1982. Frame semantics. In *Linguistics in the Morning Calm*, pages 111–138. Linguistics Society of Korea, Seoul.

Charles J. Fillmore. 1985. Frames and the semantics of understanding. *Quaderni di Semantica*, 6(2):222–254.

Charles J. Fillmore. 2014. Frames, constructions, and FrameNet. In Thomas Herbst, Hans-Jorg Schmid, and Susen Faulhaber, editors, *Constructions, Collocations Patterns*, pages 121–166. De Gruyter Mouton, Berlin.

Maucha Gamonal. 2022. A descriptive study of metaphors and frames in the multilingual shared annotation task. In *Proceedings of the Workshop on Dimensions of Meaning: Distributional and Curated Semantics*.

Luca Gilardi and Collin Baker. 2018. Learning to align across languages: Toward multilingual framenet. In *Proceedings of the Eleventh International Conference on Language Resources and Evaluation (LREC 2018)*, Paris, France. European Language Resources Association (ELRA).

Karin Kipper, Hoa Trang Dang, and Martha Palmer. 2000. Class-Based Construction of a Verb Lexicon. In *Seventeenth National Conference on Artificial Intelligence*, Austin, TX. AAAI-2000.

Ilia Kuznetsov and Iryna Gurevych. 2018. From text to lexicon: Bridging the gap between word embeddings and lexical resources. In *Proceedings of the 27th International Conference on Computational Linguistics*, pages 233–244, Santa Fe, New Mexico, USA. Association for Computational Linguistics.

Guillaume Lample, Alexis Conneau, Marc'Aurelio Ranzato, Ludovic Denoyer, and Hervé Jégou. 2018. Word translation without parallel data. In *International Conference on Learning Representations*.

Lane Lawley, Benjamin Kuehnert, and Lenhart Schubert. 2021. Learning general event schemas with episodic logic. In *Proceedings of the 1st and 2nd Workshops on Natural Logic Meets Machine Learning (NALOMA)*, pages 1–6, Groningen, the Netherlands (online). Association for Computational Linguistics.

Lane Lawley and Lenhart Schubert. 2022. Logical Story Representations via FrameNet + Semantic Parsing. In *Proceedings of the Workshop on Dimensions of Meaning: Distributional and Curated Semantics*.

Andrea Lekkas, Peter Schneider-Kamp, and Isabelle Augenstein. 2022. Multi-sense language modelling. In *Proceedings of the Workshop on Dimensions of Meaning: Distributional and Curated Semantics*.

David D. McDonald, editor. 1994. *The Balancing Act: Combining Symbolic and Statistical Approaches to Language*. Association for Computational Linguistics, Nonantum Inn, Kennebunkport, Maine.

Kyoko Ohara, Seiko Fujii, Shun Ishizaki, Toshio Ohori, Hiroaki Saito, and Ryoko Suzuki. 2004. The Japanese FrameNet Project; An Introduction. In *Proceedings of the Workshop on Building Lexical Resources from Semantically Annotated Corpora*, pages 9–12, Lisbon. LREC 2004.

Martha Palmer, Daniel Gildea, and Paul Kingsbury. 2005. The Proposition Bank: An Annotated Corpus of Semantic Roles. *Computational Linguistics*, 31(1):71–106.

Miriam R. L. Petruck. 1996. Frame Semantics. In Jef Verschueren and Jan-Ola Östman, editors, *Handbook of Pragmatics: Traditons*. John Benjamins Publishing Company, Amsterdam, Philadelphia.

Peng Qian, Tahira Naseem, Roger Levy, and Ramón Fernandez Astudillo. 2021. Structural guidance for transformer language models. In *Proceedings of the 59th Annual Meeting of the Association for Computational Linguistics and the 11th International Joint Conference on Natural Language Processing (Volume 1: Long Papers)*, pages 3735–3745, Online. Association for Computational Linguistics.

Josef Ruppenhofer, Michael Ellsworth, Miriam R. L. Petruck, Christopher R. Johnson, Collin F. Baker, and Jan Scheffczyk. 2016. *FrameNet II: Extended Theory and Practice*. International Computer Science Institute, Berkeley, California. Distributed with the FrameNet data.

Roger C. Schank and Robert P. Abelson. 1977. *Scripts, Plans, Goals and Understanding: an Inquiry into Human Knowledge Structures*. Lawrence Erlbaum, Hillsdale, NJ.

Ekaterina Shutova, Beata Beigman Klebanov, and Patricia Lichtenstein, editors. 2015. *Proceedings of the Third Workshop on Metaphor in NLP*. Association for Computational Linguistics, Denver, Colorado.

Jennifer Sikos and Sebastian Padó. 2018. Using embeddings to compare framenet frames across languages. In *Proceedings of the First Workshop on Linguistic Resources for Natural Language Processing*, pages 91–101. Association for Computational Linguistics. Collocated with COLING 2018.

Carlos Subirats. 2009. Spanish FrameNet: A frame-semantic analysis of the Spanish lexicon. In Hans Boas, editor, *Multilingual FrameNets in Computational Lexicography: Methods and Applications*, pages 135–162. Mouton de Gruyter, Berlin/New York.

Carlos Subirats-Rüggeberg and Miriam R. L. Petruck. 2003. Surprise: Spanish FrameNet! In *Proceedings of the Workshop on Frame Semantics, XVII International Congress of Linguists (CIL)*, Prague. Matfyz-press.

Tiago Timponi Torrent, Ely Edison da Laviola Matos, Ludmila Lage, Adrieli Laviola, Tatiane Tavares, Vânia Almeida, and Natália Sigiliano. 2018. Towards continuity between the lexicon and the constructicon in FrameNet Brasil. In *Constructional Approaches to Language*. John Benjamins, Amsterdam, NL.

Tony Veale. 2016. *Metaphor, a computational perspective*. Synthesis lectures on human language technologies, No. 31. Morgan and Claypool, San Rafael, California (1537 Fourth Street, San Rafael, CA 94901 USA).

Piek Vossen, Antske Fokkens, Isa Maks, and Chantal van Son. 2018. Towards an open dutch framenet lexicon and corpus. In *Proceedings of the International FameNet Workshop 2018: Multilingual FrameNets and Constructions*, Miazaki, Japan. ELRA?

Patrick Xia, Guanghui Qin, Siddharth Vashishtha, Yunmo Chen, Tongfei Chen, Chandler May, Craig Harman, Kyle Rawlins, Aaron Steven White, and Benjamin Van Durme. 2021. LOME: Large ontology multilingual extraction. In *Proceedings of the 16th Conference of the European Chapter of the Association for Computational Linguistics: System Demonstrations*, pages 149–159, Online. Association for Computational Linguistics.

Aotao Xu and Yang Xu. 2021. Chaining and the formation of spatial semantic categories in childhood. In *Proceedings of the Annual Meeting of the Cognitive Science Society*.

Liping You and Kaiying Liu. 2005. Building Chinese FrameNet database. *Natural Language Processing and Knowledge Engineering, 2005. IEEE NLP-KE '05. Proceedings of 2005 IEEE International Conference on*, pages 301–306.

Author Index